T0172798

Shadow Algorithms
Data Miner

Shadow Algorithms
Data Miner

Andrew Woo
Pierre Poulin

CRC Press
Taylor & Francis Group
Boca Raton London New York

CRC Press is an imprint of the
Taylor & Francis Group, an **informa** business

AN A K PETERS BOOK

Cover image: "City View," by Kumi Yamashita. Used by permission.

CRC Press
Taylor & Francis Group
6000 Broken Sound Parkway NW, Suite 300
Boca Raton, FL 33487-2742

First issued in paperback 2019

© 2012 by Taylor & Francis Group, LLC
CRC Press is an imprint of Taylor & Francis Group, an Informa business

No claim to original U.S. Government works

ISBN-13: 978-1-4398-8023-4 (hbk)
ISBN-13: 978-0-367-38124-0 (pbk)

Library of Congress Cataloging-in-Publication Data

Woo, Andrew, 1965-
  Shadow algorithms data miner / Andrew Woo, Pierre Poulin.
    p. cm.
  "An A K Peters Book."
  Summary: "Digital shadow generation continues to be an important aspect of visualization and visual effects in film, games, simulations, and scientific applications. This resource offers a thorough picture of the motivations, complexities, and categorized algorithms available to generate digital shadows. From general fundamentals to specific applications, it addresses "out of core" shadow algorithms and how to manage huge data sets from a shadow perspective. The book also examines the use of shadow algorithms in industrial applications, in terms of what algorithms are used and what software is applicable. "-- Provided by publisher.
  Includes bibliographical references and index.
  ISBN 978-1-4398-8023-4 (hardback)
  1. Graph algorithms. 2. Data mining. 3. Computer graphics. I. Poulin, Pierre, 1964- II. Title.

QA166.245.W66 2012
006.6'93--dc23

2011047828

Visit the Taylor & Francis Web site at
http://www.taylorandfrancis.com

and the CRC Press Web site at
http://www.crcpress.com

To Cynthia, Marcus, Eric, Alice, Carson, and Shirley

—Andrew Woo

À Thérèse et Robert

—Pierre Poulin

Contents

Preface, Motivation, and Objective

Digital shadow generation is an important aspect of visualization and visual effects in film, games, simulations, engineering, and scientific applications. When we first published our original shadow survey paper [639] more than twenty years ago, we had no idea that the topic of digital shadow generation would generate so much more research. The abundance of research since the 1990s is due to several reasons:

- The difficulty of resolving the shadowing problem well.

- The need to optimize the shadow algorithms for specific real-time applications. However, this is not an invitation to omit offline rendering needs, which, sadly, has been a problem in the last decade in terms of paper publications.

- Different performance issues due to advancement in hardware, such as the focus on GPUs (graphics processing units), multicore and SIMD architectures, a switch from reducing FLOPs to accelerate algorithms to improving locality of reference of RAM, etc.

- The emergence of different data representations, such as image-based impostors, and point-based and voxel-based primitives, which, in many cases, are used alongside the more common polygonal representation.

- The emergence of visually realistic and feasible rendering capabilities that take indirect illumination (e.g., ambient occlusion, precomputed radiance transfer) and global illumination (e.g., radiosity, Monte Carlo ray tracing) into account.

This abundance of research is reflected in the large number of papers published in the past (see Figure 1). Also note the significant activities since 2000, with a

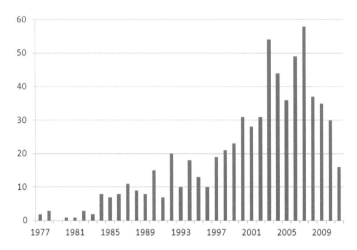

Figure 1. Number of publications published per year related to shadowing algorithms in computer graphics. Note that the paper count for 2011 is only partially available.

climb to above 30 papers published each of the following years, mainly in the real-time domain due to broad usage of the GPU. This large amount of literature results in the need to have a "shadow algorithms data miner" resource. Thus, the objective of this book is to provide an understanding of the shadow fundamentals (with little emphasis on actual code and detailed mathematical formulation) so that the reader gets an organized and structured picture of the motivations, complexities, and categorized algorithms available to generate digital shadows. The objective is further enhanced by sorting out which are the most relevant algorithms for the reader's needs, based on practitioners' experience and looking at shadow algorithms from a larger (graphics) system perspective. As a result, the reader knows where to start for his application needs, which algorithms to start considering, and which actual papers and supplemental material to consult for further details. In fact, this is why we chose the book cover from the wonderful work of Kumi Yamashita, where many numerical characters combine to cast a shadow of a woman. A data miner resource, such as this book, can clear the confusion (the many bits and bytes) and help developers to achieve results that are simple, clever, and elegant (the shadow of a woman).

Note that we have deliberately included older references in this book as well. This is important not only for acknowledging the authors in the older works, but also because there is likelihood that some older techniques can prove useful in the future due to advances in hardware and other algorithms. For example, many ray tracing acceleration papers, published back in the 1980s, have come back in the last five years due to the emergence of ray tracing for near-real-time needs. The older techniques also may be more appropriate for mobile devices due to the assumption of a lower baseline set of GPU capabilities. In fact, the offline techniques of today may become the real-time techniques of tomorrow.

Finally, in order to focus our attention, this book concentrates purely on the 3D digital generation of shadows within the domain of 3D computer graphics. There are related and important domains in which shadows play a key role, such as algorithms in computer vision which detect shadows within images, but we do not intend to delve into these algorithms unless they are directly connected to the ability to digitally generate shadows.

Before starting on the main content of this book, we first go over the organization of the book, followed by the consistent notation used throughout the book, and then the state of graphics consumer hardware.

Enjoy!

Organization

This book represents the complete published work (as far as we could find) up until the middle of 2011 for shadow generation from direct illumination approaches; indirect and global illumination approaches are only lightly touched on. However, due to the abundance of papers, we are bound to miss or misinterpret some papers—we apologize for this ahead of time.

Note that it is recommended that the reader have some background in 3D rendering before proceeding with this book.

1. If the reader has none, a good introductory book is *Fundamentals of Computer Graphics* by Shirley and Marschner, Third Edition, published by A K Peters, Ltd. [525].

2. For a more advanced coverage of the topics, please refer to *Real-time Rendering* by Akenine-Möller, Haines, and Hoffman, Third Edition, published by A K Peters, Ltd. [7].

A website is associated with this book; it can be accessed at the following address http://www.iro.umontreal.ca/~poulin/SADM. It provides additional complementary information, and it will be updated over time.

Our book is organized into major chapters, including

Chapter 1: Preliminaries of Shadows. This chapter provides basic concepts and high-level categorized approaches for shadow determination. This chapter must be read before the reader should proceed to the next chapters. The rest of the chapters are pretty much self-contained, although we recommend that the order of reading be in the order it is written. The hard shadow chapter should be read before the soft shadow chapter.

Chapter 2: Hard Shadows. This chapter provides an overview of the major approaches used to compute hard shadows, i.e., shadow boundaries with hard edges. The typical light types that generate hard shadows include the directional, point, and spot lights. After this chapter, if the reader only needs to consider polygons, then our recommendation is to skip Chapter 3 for now and proceed to Chapter 4. However, if non-polygons need to be considered, Chapter 3 will address this subject.

Chapter 3: Supporting Shadows for Other Geometry Types. The major geometry type discussed throughout this book mainly focuses on polygons (in most cases, triangles are assumed), which is a very common graphics primitive. This chapter provides an overview of the major approaches used to compute shadows for non-polygonal primitives, which can be quite different from the polygonal algorithms in some cases. Such primitives include higher-order surfaces, image-based rendering for impostors, geometry images, particle systems, point clouds, voxels, and heightfields. Most of the algorithms discussed in this chapter belong to hard shadows, as very little literature exists for soft shadow algorithms of non-polygonal objects.

Chapter 4: Soft Shadows. This chapter provides an overview of the major approaches used to compute soft shadows, i.e., shadow boundaries with soft edges. The typical light types that generate soft shadows include extended lights such as linear, polygonal, area, spherical, and volumetric lights. Other important sources of soft shadows can come from motion blur, ambient occlusion, precomputed radiance transfer, and global illumination. These will be covered in the next chapter.

Chapter 5: Other Treatments of Shadows. This chapter provides an overview of other treatments of shadows and shadow computations, such as bump mapping, advanced reflection models, semitransparency, highly complex thin materials, atmospheric shadows, motion blur, ambient occlusion, precomputed radiance transfer, and global illumination.

Chapter 6: Applications of the Shadow Algorithms. This chapter provides some insight into other applications of the shadow algorithms discussed thus far, including supporting augmented reality, non-photorealistic rendering, and using shadows as interaction tools.

Chapter 7: Conclusions. This last chapter provides some concluding remarks about shadow determination algorithms.

Notation

Unless stated otherwise, the commonly used formatted variables that are used consistently throughout this book are given in Table 1, and illustrated in Figure 2.

Variable	Definition
P	Point (x, y, z) to be shaded.
$\hat{N}$	Unit normal (x, y, z) at point P.
L	Point light source position (x, y, z), or point sample on an extended light source.
$\hat{L}$	Unit light vector from P to L; thus, $\hat{L} = \frac{L-P}{\|L-P\|}$ (note that a directional (infinite) light is only defined by $\hat{L}$).
C	Camera position (x, y, z).
$\hat{V}$	Unit view vector (x, y, z) for camera at C pointing at P; thus, $\hat{V} = \frac{P-C}{\|P-C\|}$.

Table 1. Notation used in this book.

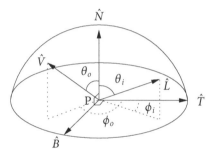

Figure 2. Illustration of the commonly used variables.

Spaces

When rendering a 3D scene in an image, as well as when computing shadows, various stages of the process can be accomplished in different spaces, most often for efficiency purposes.

When we refer to *world space*, we assume the coordinates are given according to the 3D space of the entire scene. In *object space*, the coordinates are given in the local space of the object, i.e., before the object is transformed and inserted in world space. In *camera space*, the coordinates are given in the local space of the camera pyramid, formed by the C position, the image rectangle, and sometimes, near and far clipping planes. This pyramid is sometimes called view frustum. In *normalized camera space*, the pyramid has been transformed into a box of coordinates $(\pm 1, \pm 1, \{0, -1\})$. In *light space*, the coordinates are expressed in the same space as in camera space, but they take the point light source L as the camera position C. Figure 3 illustrates these different spaces.

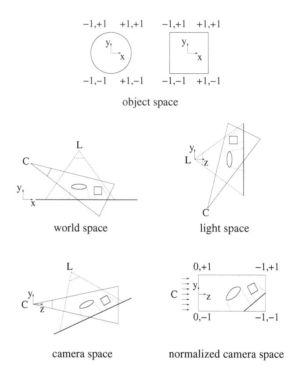

Figure 3. Different working spaces illustrated in 2D: object space, world space, light space, camera space, and normalized camera space.

State of Graphics Consumer Hardware

We feel that a snapshot of today's graphics consumer hardware provides an important context to the current algorithms described in this book. We try to present most of the older contributions in today's context, discussing their limitations when appropriate. In the future, changes in the graphics consumer hardware can significantly alter the future algorithms, as well as the effectiveness of today's algorithms. The most significant consumer hardware aspects include

Direct3D, OpenGL. The two most used APIs that can render 2D and 3D elements and take advantage of hardware acceleration (GPU) when available. Direct3D is targeted for Windows platforms, whereas OpenGL is an open industry standard that is available on many platforms. Direct3D is currently on version 11, and OpenGL is on version 4.1.

OpenGL ES. While Direct3D and OpenGL are available for PCs, OpenGL ES is the only API available on mobile devices (phones, tablets, and video game consoles). Some of the older techniques (assuming a lower baseline set of GPU capabilities) described in this book would be more suitable for mobile devices under OpenGL ES. OpenGL ES is currently on version 2.

GPU. Current mainstream GPU capabilities (available on most PCs, laptops, notebooks) include graphics capabilities such as matrix, vector, and interpolation operations, texture mapping, rendering polygons, and programmable (vertex, fragment, pixel) shaders. Current GPUs can easily support typical memory of 512 MB to 1 GB, with high-end GPUs supporting several GB. The GPU has also advanced to the point of being able to perform a lot more than just graphics operations, known as GPGPU (general purpose GPU).

CPU + GPU. An important trend has been to combine CPU + GPU on a chip (e.g., AMD's Fusion, NVIDIA's Tegra 2, Intel's Sandy Bridge, etc.). This is significant because CPU to GPU transfer (of data) becomes much less of a bottleneck.

GPGPU. There are three main APIs for GPGPU: OpenCL, DirectCompute, and CUDA. CUDA is well suited for NVIDIA-based machines, DirectCompute is well suited for DirectX-based machines, and OpenCL is attempting to target a wider set of platforms. CUDA is on version 2.1, DirectCompute is linked with DirectX-10 and DirectX-11, and OpenCL is on version 1.1.

Multicore. The PCs, laptops, and notebooks of today are minimally dualcore or quadcores (i.e., they contain two or four CPUs, respectively). Even tablets (e.g., iPad2) are dualcores. However, this is a recent trend within the last five years, so single-core machines should not ignored.

SIMD architecture. The most common SIMD architecture today would be the MMX and SSE instruction sets, where code can be written at the microprocessor level to achieve much higher speeds than writing in a standard programming language.

32-bit and 64-bit architectures. Most PCs and laptops today support 64-bit architectures, which means that the machine can support RAM sizes significantly larger than 2 GB. However, there are still existing PCs as well as supporting software that are only capable of 32-bit, which means that any application can only use 2 GB of memory (in most cases, it is actually around 1.6 GB, since memory is needed to run the OS).

Acknowledgments and Dedication

We would both like to dedicate this book to Alain Fournier, who passed away on August 8, 2000. Alain was our thesis supervisor, our teacher, our friend, and the third author in our original shadow survey paper [639]. Whenever people would refer to our paper as "Woo et al.," Alain would jokingly refer to himself as the "Al." A website dedicated to his achievements and the annual *Alain Fournier Ph.D. Dissertation Award* can be found at http://www.iro.umontreal.ca/~poulin/fournier. He is dearly missed.

This book could not have been completed without the contributions of many people, among them,

- Alice Peters, for her valuable suggestions on some topics of the book and for believing in this book despite already having another book (*Real-Time Shadows*) of a similar topic ready for publication.

- Sarah Chow, for her crucial and much appreciated help with the final stages of the book.

- Eric Haines, Mathias Paulin, Per Christensen, Torsten Möller, Shoko Leek, Derek Nowrouzezahrai, Marc Stamminger, and Eugene Fiume, for providing invaluable suggestions about our book.

- Luc Leblanc, Thomas Annen, Lee Lanier, Tom Klejne, Victor Yudin, Aghiles Kheffache, Gilles-Philippe Paillé, and Dorian Gomez, for providing some extra renderings done specifically for this book.

- Lev Gretskii, Marc Ouellette, Tomas Akenine-Möller, Frédo Durand, Christian Laforte, Luc Leblanc, Jocelyn Houle, and Kelvin Sung, for reviewing and providing valuable suggestions towards early versions of our work.

- Craig Denis, Rob Lansdale, Carl Byers, Peter Shirley, Gordon Wong, Sherif Ghali, David Salesin, Parveen Kaler, Mike Lee, Zhigang Xiang, Kyunghyun Yoon, Yung-Yu Chuang, for additional information on algorithms used in particular papers or approaches.

About This Book's Authors

Andrew Woo's experience in technology research and development spans over 20 years. In his role as NGRAIN's Chief Technology Officer, he has been responsible for applied research, product development, quality assurance, professional services, instructional design, and intellectual property for products and projects that have been deployed throughout the Canadian and US militaries. Prior to working with NGRAIN, he was senior R&D Manager at Alias-Wavefront, where he was responsible for rendering in award-winning 3D graphics products including Sketch!,

Studio, and Maya, for which the Alias team won an Academy Award in technical achievement in 2003. He has published over 20 technical papers on the subject of computer graphics. His ideas have been implemented in products from NVIDIA, Microsoft, and Pixar. Andrew is cofounder of the ACM SIGGRAPH Chapters in Toronto and Vancouver. He is an ACM Distinguished Member and is a member of IEEE, CHCCS, and CIPS. Andrew received his BSc and MSc in Computer Science and Commerce from the University of Toronto, in 1987 and 1989, respectively. Andrew also holds designations in I.S.P. and ITCP.

Pierre Poulin is a professor in the department of computer science and operations research (DIRO) at the Université de Montréal, where he has been teaching and doing research in computer graphics since 1994. He has supervised more than 50 PhD and MSc students in a wide range of computer graphic topics, in which he regularly publishes. These topics include image synthesis, real-time rendering, local and global illumination, appearance modeling, inverse shading, procedural modeling and animation, natural phenomena simulation, character animation, geometry images, geometry synthesis, and visualization. He serves on the editorial board of the *Computer Graphics Forum* and *Revue Electronique Francophone d'Informatique Graphique*, has been copresident of four international conferences, has participated in more than 45 program committees of international conferences, and has reviewed more than 400 papers in journals and conferences, and more than 40 grant proposals. He has been vice president of CHCCS since 2002.

Pierre received his BSc in 1986 from the Université Laval in Québec City, his MSc in 1989 from the University of Toronto, and his PhD in 1994 from the University of British Columbia, all in computer science. He spent a six-month postdoc at Princeton University and has been involved in various research in laboratories during his sabbaticals at Alias-Wavefront (now Autodesk), SquareUSA, INRIA Sophia Antipolis, IRIT (Paul Sabatier and ENSEEIHT), and LIRIS (Lyon 1).

CHAPTER 1
Preliminaries of Shadows

1.1   What's Covered in This Chapter

This chapter covers the basic concepts of shadows, including an introduction of the perceptual impact of hard, soft, and colored shadows (Section 1.2). It then classifies the algorithms into major categories (planar receivers of shadows, shadow depth map, shadow volume, ray tracing, etc.), and discusses the high-level algorithm as well as advantages and disadvantages of the categorized approaches (Section 1.3). Because this book mainly deals with shadowing based on occlusion of other objects onto the shadowed object, self-shadowing also needs to be considered (Section 1.4). This chapter concludes with the many considerations for choosing the appropriate shadow-determination algorithm, which are usually dependent on non-shadow factors (Section 1.5).

1.2   Basic Concepts

In the upcoming sections, the basic concepts of shadows are discussed, including why shadows are important (Section 1.2.1), the concepts of hard and soft shadows (Section 1.2.2), and the causes of colored shadows (Section 1.2.3).

1.2.1   Why Shadows

A shadow is a region in 3D space where light emitted or reflected is completely or partially occluded. As such, computing shadows is the same as computing the visibility of the light emitter or re-emitter for a region.

Using almost any measure of image quality, the computation of shadows is essential. They cause some of the highest-intensity contrasts in images; they provide strong clues about the shapes, relative positions, and surface characteristics of the objects (both occluders and receivers of shadows); they can indicate the approximate location, intensity, shape, size, and distribution of the light source; and they

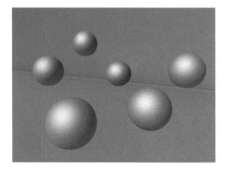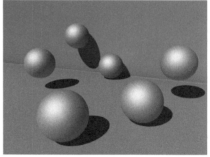

Figure 1.1. Shadows provide visual information on relative positions of the spheres on the floor and wall. *Image courtesy of Tom Klejne.*

represent an integral part of the sunlight (or lack of) effect in architecture with many buildings. Figure 1.1 illustrates clues that shadows provide in context of the same image without and with shadows.

In fact, in some circumstances, shadows constitute the only components of the scene, as in shadow-puppet theater and in pinscreen animation, developed by Alexander Alexeieff and Claire Parker [457, 365]. Another wonderful treatment of shadows comes from "shadow art," including the work of Kumi Yamashita (see the book cover), as well as Paul Pacotto [449], who uses a sculpture of a rose to cast a shadow of a woman (see Figure 1.2). An equivalent form of shadow art can be seen

Figure 1.2. Sculpture of a rose casting a shadow of a woman. *Photograph courtesy of Paul Pacotto.*

Figure 1.3. An equivalent of shadow art in computer graphics form. ©2009 ACM, Inc. *Included here by permission [414].*

in computer graphics [414], although the visual effect is purely shadows-based, where there are no aesthetics in the shadow casting model (see Figure 1.3).

Perceptual Impact of Shadows

Wanger [623] evaluates various depth cues that are useful for displaying inter-object spatial relationships in a 2D image. Shadows form an important visual cue among the depth cues. However, comparing the use of hard shadows versus soft shadows as visual cues, he determined that hard shadows are actually more beneficial as a visual cue. Studies have been done in which the availability of shadows improves the interaction of object positioning and improves the accuracy of spatial relationships between objects [622, 30, 380, 251, 375] and the perception of realism [472]. In fact, without the presence of shadows, surfaces often appear as if they are floating over a floor when they are actually lying on the floor (as can be seen in Figure 1.1). This is why shadows are one of the crucial elements in some augmented reality applications (see Section 6.2).

One conclusion to draw from these studies is that shadows form important visual cues for spatial relationships between objects and light sources in a 3D scene. However their exact determination might not be as important as long as they are consistent with our expectations. In fact, when dealing with extended light sources, exact shadows can be very surprising and unnatural for the average observer. The three images in Figure 1.4 show the shadow from a linear light aligned with the

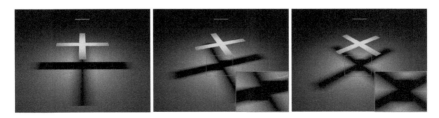

Figure 1.4. A cross-shaped object is rotated by 20 degrees and 40 degrees around the vertical axis. The soft shadow cast on the plane below from a thin elongated linear light source becomes discontinuous under rotation.

Figure 1.5. Super Mario casts a simplified circular shadow, which remains effective enough in this Nintendo *Super Mario 64* game.

object (left) and with the object rotated by 20 degrees (center) and by 40 degrees (right) around the vertical direction. Notice the (correct) discontinuity within the shadows.

This can lead to potential shadow approximations that exploit visual expectations and therefore to simplifications over the visibility algorithms. Direct image visibility is always more demanding than shadows, which are a secondary phenomenon. The next set of chapters will show a number of such approximations that are commonly used for shadows.

We can go even further. Shadows reflecting the shape of the occluder can be very approximate and remain very effective in certain real-time environments such as video games. For example, in *Super Mario 64*, Super Mario casts a circular shadow on the ground (see Figure 1.5). Although the shadow does not reflect the silhouette of (the occluder) Super Mario, it is a very effective, real-time visual cue as to where Super Mario is with respect to the ground. That such simplified shadows satisfy many users is actually confirmed by two studies [427, 499]. Further, shadow algorithms have also been introduced that are less accurate but acceptable for moving objects [402].

1.2.2 Hard Shadows versus Soft Shadows

Shadow determination, in the context of occlusion from other surfaces, can be considered some variation of the visibility determination problem. Instead of computing visibility from the camera, however, shadow determination computes visibility from the light source. One main difference is that for shadow determination, it is not necessary to calculate the closest visible surface; it is only necessary to determine if there is occlusion between the surface and the light.

Figure 1.6. Hard shadows versus soft shadows. *Image courtesy of Hasenfratz et al. [227], ©Eurographics Association 2003. Reproduced by permission of the Eurographics Association.*

There are basically two shadow types: hard and soft shadows. Figure 1.6 shows an illustration of the two shadow types. We will discuss the shadow types in detail in the subsequent chapters.

For existing shadow algorithm surveys, refer to the following important publications:

1. An older survey by Woo, Poulin, and Fournier (1990) [639].

2. A real-time, soft shadow algorithm survey by Hasenfratz, Lapierre, Holzschuch, and Sillion (2003) [227].

3. A book that includes a real-time shadow algorithm survey by Akenine-Möller, Haines, and Hoffman (2008) [7].

4. A real-time, hard shadow depth map survey by Scherzer, Wimmer, and Purgathofer (2010) [507].

5. A recent book by Eisemann, Schwartz, Assarsson, and Wimmer (2011) [158], which provides in-depth description, mathematics, and analysis of real-time shadow algorithms.

Hard Shadows

A hard shadow is the simplest type of shadow, displaying only the umbra section. If a region of space is either completely occluded or completely lit, hard-edged boundaries are formed between the shadowed (umbra) and lit regions. Calculation of hard shadows involves only the determination of whether or not a point lies in shadow of occluding objects. This is a binary decision problem on top of the illumination model. In other words, we multiply a value of either 0 or 1 by the reaching light intensity, indicating in shadow or not in shadow, respectively. The types of light sources truly generating hard shadows include a point light, spotlight, and directional light (Figure 1.7). Chapter 2 covers algorithms that generate hard shadows for polygons, and Chapter 3 does the same for non-polygonal primitives.

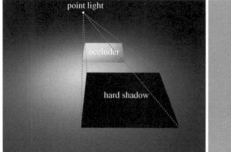

Figure 1.7. Point (left) and directional (right) light casting hard shadows.

Soft Shadows

The other type of shadow is a soft shadow, or the inclusion of a penumbra region along with the umbra for a higher level of quality resulting from an extended light source (although there can be cases of the combination of light and object positions where the umbra or the penumbra may not be visible). Full occlusion from the light causes the umbra region, and partial occlusion from the light causes the penumbra region. The degree of partial occlusion from the light results in different intensities of the penumbra region. The penumbra region causes a softer boundary between shadowed and fully lit regions. The resultant shadow region is a function of the shapes of the light source and the occluder. Instead of a binary decision on top of the illumination model as for hard shadows, a fraction in the range of [0,1] is multiplied with the light intensity, where 0 indicates umbra, 1 indicates fully lit, and all values in between indicate penumbra. Needless to say, soft shadows require more computations than hard shadows, and the soft-shadow algorithms are also more complex.

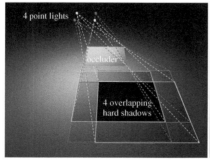
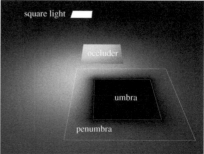

Figure 1.8. Soft shadows from area light.

The types of light sources generating soft shadows include linear, polygonal/area, and spherical lights—actually any extended light. Figure 1.8 shows an example of a soft shadow due to a square-shaped area light (right) and a poor approximation due to point lights located at the corner of the same square-shaped light (left). Soft-shadow algorithms from extended lights are discussed in detail in Chapter 4.

Different regions within hard and soft shadows can be observed. Each such region corresponds to different gradients of the visibility function of the light source. These regions are defined by visual events generated from combinations of vertices and edges from the light source and occluders. The theory behind these visual events has been studied for polygonal scenes, and is covered in Section 4.3.

There are other sources of soft shadows, which include

- *Motion blur*, which is critical for film and video (see Section 5.7).

- *Ambient occlusion*, which fakes skylight with the look of soft shadows as if all objects are under an overcast day (see Section 5.8).

- *Precomputed radiance transfer*, which is mainly effective in real-time, diffuse, low-frequency lighting environments, assuming infinitely distant lights (see Section 5.9).

- *Global illumination*, which includes radiosity and Monte Carlo ray tracing techniques (see Section 5.10).

1.2.3   Colored Shadows

One assumption is that a shadow can only appear as a black/grey region. This is not always the case, especially with multiple colored lights that can result in colored shadows.

Figure 1.9.   Colored shadows from three (red, green, blue) colored lights. ©*1993 IEEE. Reprinted, with permission, from [643].*

Figure 1.10. Photograph of colored shadows from semitransparent objects.

Take the example of a blue light and a red light shining on a white-ish object. A region occluded only from the blue light will appear red, and a region occluded only from the red light will appear blue. Only a region occluded from both lights will appear black. In the case of three colored lights, when one color of light is blocked with an object, the color of its shadow is the sum of the two remaining colors. See a visual example of three colored (red, green, blue) lights in Figure 1.9.

Another form of colored shadows can come from occlusion of semitransparent surfaces. Instead of totally blocking the light, the light is transmitted, filtered, and altered through the semitransparent surface and lands on the receiver with some color as a result of the light transmission. The same phenomenon occurs with refractive surfaces; however, light rays deviate according to changes in the media refractive indices and light wavelengths. A visual example is shown in Figure 1.10. Dealing with such shadows is much more involved and is discussed in Section 5.4.

1.3   Shadow Algorithms: Basic Ideas and Properties

In the upcoming sections, we will review the major classes of algorithms available for shadow generation. This is to ensure that the reader has some basic understanding of these algorithms, before delving into the details in the upcoming chapters. In fact, based on the information provided here, the reader may choose to ignore certain classes of algorithms due to the high-level descriptions because certain algorithms clearly do not fit his needs. The major classes of shadow algorithms include planar receivers of shadows, shadow depth map, shadow volume, ray tracing, and area subdivision and preprocessing.

1.3.1   Planar Receivers of Shadows

In simpler applications of real-time shadows, certain assumptions may be made about the environment. One such example is that hard shadows are projected only on a planar floor or wall [57], i.e., the floor or wall does not self-shadow, and shadows from other objects in the scene are only accounted for on the floor or wall. The floor or wall is generally infinite and perpendicular to some fixed orientation, the easiest being one of the x, y, z world axes; thus, a single transformation matrix resembling oblique or perspective screen projection matrices is all that is necessary

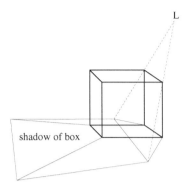

Figure 1.11. Projected shadow polygons on a planar floor.

to project the polygons' vertices onto the floor. In fact, the projected shadows can even be modeled as dark polygons, so that little additional rendering code is necessary. In Figure 1.11, vertices of the box project to vertices on the floor to determine the shadows.

Variants of this algorithm have been used in many real-time implementations when shadows on floors are needed, due to the algorithm's simplicity and speed. Hard shadow variants [567, 223] are discussed in Section 2.2, and soft shadow variants [240, 232, 550, 220, 156, 157] are discussed in Section 4.4.

1.3.2 Shadow Depth Map

In the literature, the terms shadow map and shadow buffer are often used in an inconsistent fashion. For example, a shadow buffer used in the context of voxels may refer to a 2D or 3D shadow buffer. In the context of this book, the term *shadow depth map* will be used explicitly to indicate a 2D shadow map that stores a single depth value per pixel. Any shadow map or shadow buffer references will indicate explicitly what information is stored per element and in how many dimensions it is stored. The reader may also see references in the literature to "image-based approaches" that refer to shadow depth maps in the context of shadow determination, versus "geometry-based approaches," which belong to other approaches discussed in this book, such as shadow volumes and ray tracing (to be discussed in the next sections). In general, image-based approaches have the advantage of performance while the geometry-based approaches have the advantage of accuracy.

Williams [630] uses a Z-buffer approach to determine visibility and depth with respect to the camera, and this process is repeated for the light source. Thus, like the preprocess seen in Figure 1.12, the approach creates a buffer with respect to the viewpoint of the light source L except that the buffer contains the smallest (Z_n) Z-depth values and not shading values or object information. During rendering of the camera view, each point P to be shaded is projected towards the light and

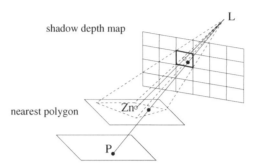

Figure 1.12. Basics of the shadow depth map algorithm; if $\|P-L\| > Z_n$, then P is in shadow.

intersects the shadow depth map pixel. If the distance $\|P - L\|$ is larger than the Z-depth value Z_n (as shown in Figure 1.12) from the projected shadow depth map pixel, then P lies in shadow; otherwise, it is fully lit. The papers by Williams [630] and Reeves et al. [476] are the most often cited and implemented versions of the shadow depth map approach.

From a GPU-based implementation, Segal et al. [515] and Everitt et al. [166] use a GPU texture (typically 16-bit, 24-bit, or 32-bit texture) to represent a shadow depth map, projectively texture it onto the scene, and then compare the depth values in the texture during fragment shading to achieve per-pixel shadows. Note that the use of the GPU texture to achieve depth map shadows may be a bottleneck on older GPUs (1990s) or lower-end platforms. An early example of rendering using the GPU is shown in Figure 1.13.

Figure 1.13. Rendering generated with a GPU-based shadow depth map. *Image from Everitt et al. [166], courtesy of NVIDIA Corporation.*

The shadow depth map has been very successful in many graphics environments because

- The basic approach is simple to implement.

- It can handle surfaces other than polygons. See Chapter 3.

- It can be used as a sort of protocol of shadow information from different geometry representations or different renderers.

- It can be used as a sort of protocol of shadow information to account for shadows from different geometry layers (that are merged into one single scene) [344].

- The performance can be quite fast without GPU assistance. Occlusion culling algorithms [106, 201, 52] can also be applied so that the light source rendering can achieve good performance while handling large data sets. The shadow depth map can handle large data sets much easier than other approaches because it only needs to process and store a single buffer per shadow casting light. Additional performance considerations are discussed in Section 2.3.2.

- It can be simply implemented in the GPU as a hardware texture [515, 166, 438, 65, 259, 69] for real-time applications, including as a standard feature in real-time engines such as Second Life, Unity 3D, DX Studio, Renderware, etc.

- Its quality is good enough that it has been used in film since the 1980s [476]. Major software renderers use some variation of the shadow depth map, such as Renderman, Maya, Mental Ray, Lightwave, etc.

- Soft shadows from extended lights have been developed based on the shadow depth map approach [570, 2, 237, 66, 652, 254, 267, 309, 268, 646, 83, 555, 16, 17, 296, 597, 173, 338, 61, 79, 489, 208, 29, 41, 27, 40, 336, 511, 209, 252, 12, 535, 43, 513, 512, 276, 486, 514, 651, 506, 417, 210, 138, 425, 522, 672]. This is discussed in Section 4.5.

- Although the basic shadow depth map approach can only deal with shadows from opaque objects, there are successful variations to handling semitransparent object shadowing as well [362, 304, 115, 555, 159, 495, 394]. This is discussed in Section 5.4.2.

- Extensions of the shadow depth map approach are the standard for handling highly complex thin materials such as hair and fur [339, 313, 362, 303, 304, 316, 405, 424, 47, 533, 655, 534, 265]. See Section 5.5.

- Extensions of the shadow depth map approach have been successful for producing atmospheric/volumetric effects [134, 135, 412, 257, 647, 186, 590, 162, 34, 90, 394]. See Section 5.6.2.

- Extensions of the shadow depth map approach can achieve soft/blurred shadows due to motion blur [548, 362]. See Section 5.7.

The disadvantages include

- Shadow determination is more complex for point lights (when placed within the convex hull of the scene) because it requires more than one shadow depth map per light [68, 307, 190, 441, 107]. This is also true for spotlights with a large angle of view because an angle of view larger than 90 degrees will likely result in poor-quality renderings if only one shadow depth map is generated. This is discussed in Section 2.3.1.

- Rendering quality issues relating to filtering and self-shadowing issues [250, 476, 642, 616, 576, 600, 298, 261, 129, 606, 166, 167, 65, 627, 75, 450, 15, 509, 596, 671, 140, 510, 11, 300, 334, 13, 43, 117, 373, 496, 302, 211] have not been completely resolved. These topics are discussed in Sections 2.3.4, 2.3.5, and 2.3.6. However, the terminator problem is avoided when using some of the self-shadowing techniques.

- The rendering quality is particularly poor when the view focuses on a specific region that only covers a small part of the shadow depth map. Many algorithms [172, 575, 557, 69, 97, 518, 387, 95, 386, 317, 4, 20, 74, 275, 633, 341, 508, 663, 664, 96, 665, 358, 632, 197, 196, 161, 214, 342, 21, 667, 670, 131, 598, 35, 359, 408, 360, 535, 599, 451, 668, 335] manage to reduce such poor results, but these algorithms can get quite complex. This is discussed in Section 2.3.3.

- Changes in the shadow coverage region can result in changes in rendering quality. By shadow coverage, we mean the world space region represented by the shadow depth map. Changes in the shadow coverage may be needed to get the sharpest image quality by encompassing only particular objects, and the particular objects' occupied world space changes during an animation. This is discussed in Section 2.3.3.

1.3.3   Shadow Volumes

Crow [114] creates shadow polygons projected from the original polygons in the opposite direction of the light and then places them into the rendering data structure as invisible polygons. The original set of polygons is also included in this rendering data structure for shadow determination and are sometimes called light caps. To compute shadow determination, a shadow count is used. An initial shadow count is calculated by counting the number of shadow polygons that contain the viewing position. The shadow count is then incremented by 1 whenever a

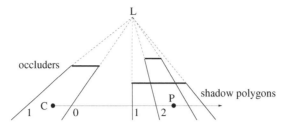

Figure 1.14. Basics of shadow volumes. Computing the shadow count for point to be shaded P from camera C.

front-facing shadow polygon (that is, entering the shadow umbra) crosses in front of the nearest visible surface. The shadow count is decremented by 1 whenever a back-facing shadow polygon (that is, exiting the shadow umbra) crosses in front of the nearest visible surface. If the final shadow count is 0, then the visible surface does not lie in shadow; if positive, it is in shadow; if negative, then it is time for a peer code review. In Figure 1.14, the initial shadow count is 1. It gets decremented/incremented to 0, 1, 2 until it hits the point to be shaded P. Since the final shadow count is greater than 0 (it is 2), then P is in shadow.

To implement shadow volumes on the GPU, Heidmann [236] draws the scene polygons shaded only by an ambient light with a hardware Z-buffer. Front-facing shadow polygons are then drawn (using a front-facing culling test), incrementing shadow counts in an 8-bit GPU stencil buffer if visible for each affected pixel. Similarly, visible back-facing shadow polygons decrement their respective shadow counts. Finally, the scene polygons are drawn with diffuse and specular shading only where their stencil shadow count is 0.

The shadow volume approach has been successfully used in some real-time graphics environments because

- It is computed at object precision and is omnidirectional, i.e., can handle shadows in any direction.

- It can very effectively produce atmospheric/volumetric effects [390, 431, 638, 152, 264, 51, 648, 49]. See Section 5.6.1.

- Soft shadow variations have been developed [71, 92, 128, 636, 595, 5, 23, 25, 24, 346, 175, 176, 177]. This is discussed in Section 4.6.

- It was one of the first GPU-supported shadow techniques and usually employs a hardware stencil buffer [236, 301, 37, 50, 81, 393, 165, 48, 5, 345, 487, 24, 23, 25, 26, 67, 397, 168, 3, 84, 357, 398, 249, 330]. In fact, GPU shadow volumes have been deployed successfully in a number of video games such as Doom 3 [601], and software such as RealityServer, EON Studio, and Sketchup. This is discussed in Section 2.4.2.

- Near-real-time variations without the need for the GPU have been developed [91, 92, 99, 100, 636, 356]. This is discussed in Section 2.4.4 and Section 4.6.2.

- Partial regeneration of shadow polygons is easy to achieve over an animation since only shadow polygons for changed objects need to be regenerated. However, this is complicated by the many optimizations needed for a performance-efficient implementation, as discussed in Section 2.4.3.

The disadvantages include

- It is primarily effective for polygonal representations, although there are more advanced variations that can deal with other geometric representations [266, 235, 580, 579, 170]. See Section 3.2.3.

- For optimal use, it needs well-formed closed objects (2-manifold shadow casters, which means an edge is shared by exactly two polygons) with adjacency information to optimize silhouette detection, although there are more generic variations [46, 8, 400, 306, 536]. This is discussed in Section 2.4.3.

- It exhibits linear growth in complexity in terms of performance, which makes it a lot less desirable in terms of larger data sets. Some optimizations are discussed in Section 2.4.3.

- Many (up to one quadrilateral per edge per light source) long shadow polygons need to be scan-converted (high fill rate). Conservative occlusion culling algorithms [106] can be applied to significantly reduce the number of shadow volumes required. Additional optimizations are discussed in Section 2.4.3; however, the optimizations discussed have not been good enough to guarantee a consistent frame rate for many real-time applications (such as many modern-day games).

- It has a limited representation of 8 to 16 bits for hardware-based shadow counts, but is most commonly just 8 bits. This issue remains unsolved and poses a problem if the shadow count surpasses 255.

- Aliasing errors exist in the shadow counts due to scan-conversion of very narrow shadow polygons. This issue remains unsolved.

- There is no obvious workaround for the terminator problem (see Section 1.4.1).

- Semitransparent objects cannot easily receive shadows when this algorithm is implemented on the GPU. The problem is that a pixel on the screen has the shadow state of only one surface, normally the closest opaque surface, stored for it. There is no additional storage for semitransparent objects that cover the pixel. Partial solutions for semitransparent-object shadowing [228, 306, 177, 536] are discussed in Section 5.4.3.

1.3.4   Ray Tracing

Ray tracing is a powerful way to render objects from the camera as well as reflections, refractions, and shadows [629]. Shadow determination using ray tracing is trivial: a shadow ray is shot from the point to be shaded P towards the light source L. If the shadow ray intersects any object between P and L, then it lies in shadow; otherwise, it is fully lit. Figure 1.15 shows a visual representation of this algorithm, and a rendering is depicted in Figure 1.16.

Ray tracing is so flexible that it is available in just about all offline rendering products because

- It is computed at object precision and is omnidirectional.

- An offset workaround is available to reduce the terminator problem (see Section 1.4.1).

- It supports surfaces other than polygons in the integration of shadow information from different geometry representations. See Chapter 3.

- There are many algorithms for soft shadow generation [108, 10, 231, 407, 465, 527, 621, 570, 463, 603, 643, 577, 33, 269, 528, 578, 452, 352, 189, 178, 297, 225, 2, 372, 328, 148, 473, 447, 64, 154]. See Section 4.7.

- Shadows from semitransparency [221, 340, 524, 454] can be achieved easily (see Section 5.4.1).

- Shadows from motion blur [108] can be achieved easily (see Section 5.7).

The primary weakness of ray tracing has been its slow performance. However, due to its flexibility and simplicity, there have been quite a number of papers [198, 526] written on the topic of accelerating ray tracing, including shadow computations. See Section 2.5 for further performance discussions [10, 231, 219, 494, 164, 638, 625, 527, 641, 94, 188, 643, 528, 171, 133]. Furthermore, in the last

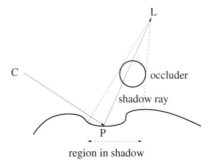

Figure 1.15. Basics of the ray tracing shadow algorithm.

Figure 1.16. Ray traced rendering with shadows. ©*1986 IEEE. Reprinted, with permission, from [219].*

decade, platform changes (SIMD instructions) and algorithmic improvements (ray packing) have allowed ray tracing to be much faster, although it is still generally not capable of being used for real-time applications unless there is sufficient hardware support for high parallelism on the CPU (multicores) [610, 613, 64] or GPU [469, 248, 213, 464, 676]. See Section 2.5.2 for more details.

Keep in mind that just because ray tracing reflections and refractions might be needed for the specific application, it is not necessarily a forgone conclusion that shadow determination must apply ray tracing as well. Shadow depth map or shadow volume algorithms can still be applied without complication (except for warping shadow depth maps (see Section 2.3.7)). However, note that most ray tracing algorithms do not require any per-light preprocessing, which is an advantage over shadow volume or shadow depth map algorithms if the number of (shadow casting) lights significantly increases.

1.3.5 Area Subdivision and Preprocessing

Nishita and Nakamae [428] and Atherton et al. [28] use clipping transformations for polygon shadow generation. In this two-pass hidden surface algorithm, the first pass transforms the image to the view of the light source and separates shadowed and lit portions of the polygons via a hidden surface polygon clipper (see Figure 1.17). It then creates a new set of polygons, each marked as either completely in shadow or completely lit. The second pass encompasses visibility determination from the camera and shading of the polygons, taking into account their shadow flag.

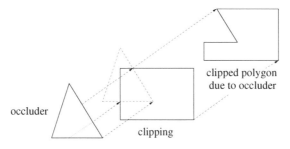

Figure 1.17. Shadow polygons clipped.

This category of shadow algorithms has received very little research attention since the original papers. Reasons for the reduced focus on this class of shadow algorithms are likely the significant increased complexity with medium-to-large data sets as well as potential numerical instability issues [183]; particularly difficult is a GPU-based implementation. Ghali et al. [191, 192] take the subdivision approach and store the shadow edge and adjacency information in a visibility map, which avoids polygon clipping instability issues, but does not resolve the other issues. Such edges can also store penumbra information for extended lights; however, a practical algorithm to compute the visibility map is needed, which makes this extended subdivision approach more theoretical so far.

Appel [14] and Bouknight and Kelly [63] generate shadows during the display using an extended scanline approach. During preprocessing of each polygon, all polygons that lie between the light source and the polygon itself are identified and stored in a list. During the display phase, polygonal boundaries from the currently scanned polygon's list are projected down onto the currently scanned polygon to form shadow boundaries, clipped within the boundaries of the currently scanned polygon, and then projected onto the viewing screen. The intensity of a scanned segment changes as it crosses the shadow boundaries. See Figure 1.18. It is easy to

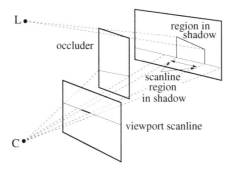

Figure 1.18. Scanline approach for calculating shadows.

see that the complexity of this approach will grow significantly with medium-to-large data sets.

Due to the lack of research papers and limited use of the above algorithms, this category of shadow algorithms will not be further discussed in this book.

1.4   Self-Shadowing

From a computer graphics standpoint, there are typically two main components for determining the shadowing: shadows due to occlusion from other objects and self-shadows. Most of the discussions in this book focus on occlusion from other objects. However, we do want to cover a few essential topics on self-shadowing to allow a more complete understanding of shadowing. Most self-shadowing issues assume the simulation of non-smooth surfaces and relate to (or are considered part of) the illumination computations or the reflection models employed, such as the $\hat{N} \cdot \hat{L}$ check, bump mapping, and advanced reflection models.

1.4.1 $\hat{N} \cdot \hat{L}$ Check

The simplest example of self-shadowing can be done using a dot product check, $\hat{N} \cdot \hat{L}$, where $\hat{N}$ is the surface normal and $\hat{L}$ is the light direction with respect to the point to be shaded. This check is done in almost all rendering systems. This means that no light directly reaches the portion of the surface that is facing away from the light without further computations. This also means that direct shading computation and shadowing from other occluding surfaces are only checked when $\hat{N} \cdot \hat{L} > 0$. This is a concept similar to back-face culling from the view direction, except it applies to the lighting direction in this case. While this check is physically correct, natural, and optimal, there are consequences to this check that should be understood, such as specular highlight cutoff and the terminator problem, which are discussed below.

Specular Cutoff

The first consequence comes from bad specular highlight cutoff [640]. Because the $\hat{N} \cdot \hat{L}$ evaluation also happens to be the diffuse reflection amount and the specular component is calculated independently of the diffuse evaluation, there can be cases where $\hat{N} \cdot \hat{L} < 0$, but the specular component is positive, indicating a specular contribution when the diffuse component has no contribution. Thus, the self-shadowing check appears to have prematurely cut off the specular component (Figure 1.19). This problem is not usually visible due to the unusual circumstances required to encounter this situation.

Terminator Problem

The second consequence comes from the terminator problem. This problem results from improper self-shadowing due to polygonal mesh approximation of a

Figure 1.19. Specular cutoff: the specular reflection component spreads where direct light should not reach because the $\hat{N} \cdot \hat{L}$ check was not performed.

smooth surface. In Figure 1.20(left), polygons A and B represent polygonal approximations to the smooth surface. At point P on A, the vertex-interpolated normal $\hat{N}'$ is used to compute the illumination as opposed to the plane's normal $\hat{N}$. Since $\hat{N}' \cdot \hat{L} > 0$, light contribution is present, and the shadow occlusion from other surfaces must be computed to determine whether P is shadowed. The shadow ray for point P intersects B and incorrectly concludes that P is in self-shadow. This artifact is usually visible as shadow staircasing and is illustrated in Figure 1.20(right), where the staircasing occurs between the dark and lit regions. A simple solution [546] is to offset the shadow ray origin by a small amount along $\hat{N}'$ to avoid the self-shadowing. Unfortunately, the correct offset value is difficult to figure out, and this offset typically assumes convex region behavior because a

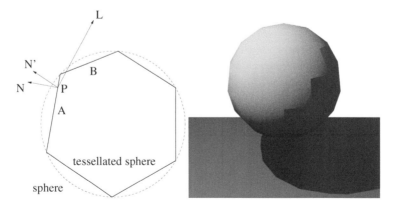

Figure 1.20. Shadow terminator problem: interpolated normal on the polygonized sphere (left); resulting shadow staircasing (right).

concave region should ideally have a negative offset. Furthermore, although this problem has been described in the context of ray tracing, it is actually a problem in all shadow algorithms, but less so in the shadow depth map and ray tracing algorithms due to the use of bias or offset factor. Further, see Section 2.3.6 for how the various shadow depth map algorithms exhibit the terminator problem. Unfortunately, a workaround has not been made available in the shadow volume approach.

A recent search on the Internet indicates that efforts to improve the terminator problem, without the need for an offset, have been attempted by the Thea render on Blender. Too few details about these attempts are publicly available, although we suspect the solution deals with identification of the silhouette to perform special-case computations and avoid the terminator problem.

1.4.2   Bump Mapping

Another example of the self-shadowing problem is bump mapping [54], where surface normals are perturbed to give the impression of a displaced, non-smooth surface. Bump mapping does not actually displace the geometry (as in displacement mapping [110]). As a result, shadowing for bump mapped surfaces appears as if the surface is perfectly smooth, because shadow determination does not use the perturbed surface normal information at all.

Techniques such as horizon mapping [391] are used to take into account the self-shadowing effects. See Figure 1.21 as an example of bump maps with and without proper self-shadowing. Please see Section 5.2 for more details on bump-map self-shadowing [391, 435, 538, 294, 238, 179, 439] as well as some advanced bump mapping effects [312, 619, 620, 396, 626, 581, 85]. Also note that some of these techniques have been useful for shadow determination of heightfields (see Section 3.8).

Figure 1.21. Bump-mapped surface without (left) and with self-shadowing (center and right). *Image courtesy of Sloan and Cohen [538], ©Eurographics Association 2000. Reproduced by permission of the Eurographics Association.*

1.4.3 Advanced Reflection Models

Direct shading is computed only when the surface is facing light ($\hat{N} \cdot \hat{L} > 0$) and is not in shadow. However, even under direct illumination, shadowing can occur within the reflection model itself.

When the bumps are smaller and denser over a surface, such that many bumps fit within a pixel, neither the bumps nor the shadows can be perceived. However, the reflection of light behaves differently as bumps cast shadows on some areas and not on others. This phenomenon should be captured by the local reflection model. In advanced reflection models, such as anisotropic reflection models, proper self-shadowing is needed for the correct visual effect. Section 5.3 discusses details on self-shadowing computations needed for advanced reflection models, and related self-shadowing computations with respect to highly complex thin materials can be seen in Section 5.5.

1.5 Considerations for Choosing an Algorithm

The choices of a shadow algorithm will get more and more complex because the following issues need to be considered while the reader is going over the algorithms described in the next set of chapters. Note that some considerations are part of the system one may be trying to build, external to the shadow algorithms themselves. Such considerations are addressed by questions like

- Are real-time or interactive speeds a requirement, and is real-time feedback from dynamic scenes a requirement?

- What is the requirement to handling data complexity (e.g., large data sets)? The appropriateness of certain classes of algorithms to handle large data sets is discussed in some of the Trends and Analysis sections.

- What platform dependencies or constraints exist? On certain older or smaller platforms, a decent GPU may not be available, so one may need to resort to efficient CPU-based algorithms or assume fewer GPU capabilities. On current tablets (e.g., iPad, Android, PlayBook), memory available to the applications is quite limited and may require out-of-core techniques (Sections 2.5.2, 2.6, 3.7.6). Also, the opportunity for extreme parallelism (multicore) can influence the choice of algorithms, particularly affecting ray tracing approaches (see Section 2.7 for further details).

- If the GPU is being applied, care must be taken if IP protection of the geometry (being displayed) is of critical importance. There are techniques to intercept geometry information on the GPU, such that the act of viewing the information could result in IP theft of the geometry. This is made worse by shadow algorithms because the shadow computations provide another view of the geometry.

Feature	Shadow Map	Shadow Volumes	Ray Tracing
Hard shadows	✓	✓	✓
Higher-order surfaces	✓	✓	✓
Point clouds	✓		✓
Voxels	✓		✓
Heightfields			✓
Soft shadows	✓	✓	✓
Semitransparency	✓	✓	✓
Complex thin materials	✓		
Atmospheric shadows	✓	✓	✓
Motion blur	✓		✓
Global illumination	✓		✓

Table 1.1. Overview of shadow algorithm features.

- What is considered a sufficient quality or accuracy level for your needs? In particular, for offline renderings, the antialiasing quality usually needs to be much higher, not only spatially but also temporally in certain domains. Especially for production needs, one can pose the question of whether it is better to have an algorithm be 80% correct, or always consistently wrong? Unfortunately for temporal aliasing, 80% correct can be quite damaging because it may result in flickering, which is very disturbing to the human visual system, whereas the consistently wrong may not be noticed by the human visual system. While this is not a true measurement of how good an algorithm is, and the above question is skewed by "how wrong the result is," it does make us wonder how stringent an algorithm might need to be to meet offline rendering needs.

- What features from the shadow algorithms are needed? See Section 1.3 of this chapter for the capabilities inherent within each of the major approaches as well as Table 1.1 for a high-level feature overview.

- Are there any constraints on the type of geometry used for rendering or interaction purposes? See Chapter 3 for shadow algorithms for other (than polygons) geometry types, including higher-order surfaces, image-based impostors, geometry images, particle systems, point clouds, voxels, and heightfields.

- Are there dependencies on the visibility determination algorithms used? In some cases, reusability of the code for visibility determination and then for shadow calculations is an attractive feature, not only because it reduces the initial coding effort, but it also decreases future code maintenance. This may be an important consideration for visibility approaches such as ray tracing (for visibility and shadow rays) and Z-buffer (for shadow depth maps).

- Similarly, are there opportunities for reduced or shared code maintenance when combining different features of shadows (e.g., hard shadows, soft shadows, semitransparent shadows, other applications of shadow algorithms, etc.)? In some of the Trends and Analysis sections, code maintenance is discussed as one of the factors in the overall decision of algorithm choice.

- How much user intervention and user input are acceptable? Offline renderings tend to be a lot more patient with user interventions, in the hunt for good-quality results, whereas real-time renderings usually assume more automatic approaches and less user input.

- Is the algorithm you wish to implement within a product already patented? If so, it is time to talk to legal council and discuss whether it is appropriate (businesswise) to base an implementation on top of a patented solution. We were originally going to list all the patents (that we know of) on shadow algorithms as part of this discussion because we thought this would be a useful service to the reader. However, in better understanding the patent policy of certain companies, this service may turn into a disservice. As a result, we have removed such a list here, and the references to patents and their implications are entirely removed from this book. If the reader is interested in such a patent list as well as how patents have affected the literature and adoption of these algorithms, we invite the reader to contact us directly for an addendum.

The above considerations in conjunction with the Trends and Analysis sections of the following chapters can help determine which category of algorithms is most appropriate for the reader's current needs.

CHAPTER 2
Hard Shadows

2.1   What's Covered in This Chapter

This chapter covers algorithms that generate hard shadows. The discussions around each categorized approach include details of individual papers and details on how to make each approach feasible in terms of image quality and performance. The main categorized approaches include

- Fake shadows on planar receivers (Section 2.2) [7, 355].

- Shadow depth maps (Section 2.3) [7, 354, 507, 158].

- Shadow volumes (Section 2.4) [7, 355, 158].

- Ray tracing (Section 2.5) [613].

The chapter concludes with other miscellaneous hard shadow algorithms (Section 2.6), followed by trends and analysis of the criteria for application, success, and deployment of each categorized approach (Section 2.7).

2.2   Planar Receivers of Shadows

Blinn [57] provides the basic mathematics and CPU implementation details needed for the projection of fake shadows onto a planar floor or wall, using a set of black polygons to represent the shadows. To make sure that the shadows show up, an offset is created so that the projected shadow polygons are visible over the floor or wall. While this is somewhat effective, it is tricky to come up with a good offset value that is artifact free. Instead, Hallingstad [223] implements the fake shadows on a plane on the GPU using a stencil buffer. This is done by first rendering the scene without shadows, then during the shadow casting (projection) step, the depth test is disabled and blending enabled for the floor. For static scenes, it is

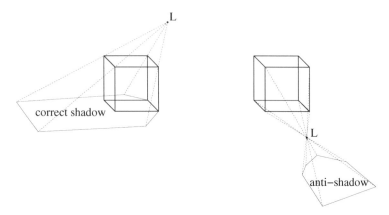

Figure 2.1. Correct shadowing (left). Example of anti-shadows (right).

also interesting to calculate the shadow polygons and place the shadowing onto a texture map for that planar receiver.

One limitation to this approach is that this floor or wall does not cast shadows, it just receives shadows. Also, the common assumption is of an infinite floor or wall so that no clipping of the shadow polygons are needed, although there is no reason (except the performance hit) not to do the clipping in the presence of a finite floor. Another common procedure is to only project front-facing polygons.

Care must be taken to avoid anti-shadows. Anti-shadows occur when a point light is located between the planar receiver and the occluder, and shadows actually show up on the receiver incorrectly. This is not a concern if the light is outside the convex hull of the scene (which is the assumption made in certain environments [567]). Anti-shadows can usually be detected by doing dot product tests with

Figure 2.2. Fake shadows on planar receivers. Note that shadows are only cast on the floor and not on the other objects. *Image courtesy of Transmagic Inc.*

respect to bounding boxes, but this method is not foolproof. See Figure 2.1(left) for correct shadowing and Figure 2.1(right) for an example of anti-shadows.

An example of shadows generated in this manner is given in Figure 2.2. Note that shadows only exist on the floor and not on the object itself. This is a very common occurrence in real-time applications, where only the shadows on the floor serve as good points of reference. However, (fake) soft shadows are often desirable, which is why Transmagic uses a simplified shadow depth map approach (Section 2.3), as seen in Figure 2.2, instead of the Blinn approach. The shadow depth map is simplified because this particular use does not require it to be concerned with self-shadowing (Section 2.3.4; this problem is avoided because the floor does not need to be part of the shadow depth map, and thus self-shadowing on the floor is never checked), and the uniform blur operation can simply be handled using percentage-closer filtering (Section 2.3.5).

2.3   Shadow Depth Maps

While the shadow depth map appears simple, one must consider many aspects when efficiently and robustly rendering a data set. Among them are how to deal with different light types (Section 2.3.1) and the impact of these shadows on performance (Section 2.3.2) and quality (Section 2.3.3).

2.3.1   Dealing with Different Light Types

For the shadow preprocess rendering, it is important to realize that a spotlight maps to a perspective rendering from the light. However, the circular region often represented by the spotlight must be totally inside the perspective rectangular region, resulting in some wasted rendering region. Similarly, a directional light maps to an orthographic rendering from the light, but the bounding region of the rendering must be clipped with the scene (or portion of the scene that is visible to the camera).

Things get a bit more complicated when dealing with point lights because a single perspective rendering cannot cover a view frustum of 360 degrees. For a point light source, multiple shadow depth maps need to be generated using a perspective rendering. The most common approach produces six 90-degree-view shadow depth maps, such that the six views form a full cube around the point light. In this way, full coverage in both the azimuth and altitude of the point light is considered for shadow computations. During the actual shading phase, the point to be shaded is projected on one of the six shadow depth maps to determine the shadow occlusion. An implementation of the above can be achieved using cube-maps [307], which have also been referred to as omnidirectional shadow depth maps. Gerasimov [190] goes over the code details for omnidirectional shadow depth maps. This cube-map structure is fairly standard on the GPU today.

Note that the above method may result in discontinuous artifacts (when zoomed in) when the point to be shaded projects close to the border of one of the six shadow depth maps. This is due to the perspectives changing drastically at the borders when moving from one shadow depth map to another. A simple workaround is to slightly overlap the six shadow depth maps, computing six 95-degree-view shadow depth maps instead of 90 degrees. When a point projects to any of the overlapping regions, then the shadow lookup is done for multiple shadow depth maps (up to a maximum of three), and their results are averaged. This alleviates most of the discontinuous artifacts, although it is slower in performance.

Note that six such shadow depth maps are not always required for point lights. In many cases in which the point light is above all surfaces (e.g., the light is on the ceiling), the number of shadow depth maps can be reduced or, put another way, if this optimization is not done, some shadow depth maps may be empty.

Brabec et al. [68] deal with the point-light shadow depth map problem by employing a dual-paraboloid mapping. Instead of six planar shadow depth maps forming a full cube, only two hemispherical shadow depth maps are needed. Each paraboloid map can be seen as an image obtained by an orthographic camera viewing a perfectly reflecting paraboloid. However, with large polygons, linear interpolation performed during rasterization could cause nonlinear behaviors. As well, it is likely that some discontinuity artifacts might appear near the hemisphere borders. Implementation details of this approach are documented by Osman et al. [441], and the nonlinear behavior is bypassed by tessellating the surfaces based on the distance from the light (i.e., higher tessellation is applied when closer to the light).

Contreras et al. [107] propose an approach to perform only a single render pass by using dual-sphere-unfolding parameterization. Both the dual-paraboloid and dual-sphere-unfolding approaches remain more experimental at this point, as standard approaches to handle self-shadowing properly (Section 2.3.4) require more examination, and it is difficult to incorporate them with some focus-resolution approaches (Section 2.3.7), especially nonlinear shadow depth maps.

2.3.2   Performance Considerations

There are a number of performance considerations to optimize the performance of a shadow depth map implementation, such as faster shadow depth map generation, temporal coherence considerations, and memory issues.

General Performance Techniques

Zhang [669] introduces the concept of forward shadow mapping to exploit spatial coherence in the shadow depth map when rendering. The shadow depth map is computed as usual. This is followed by rendering from the camera view without accounting for shadowing. Each shadow depth value per pixel is then warped to the camera view. The depth value of each camera view pixel is compared against the warped shadow depth value. If they are approximately the same, the camera view pixel is lit. The lit/non-lit status is put into a modulation image. The modulation image is blurred to achieve antialiasing of the shadow edges. This has the advantage that the shadow depth map pixels, as well as the camera-view pixel values, are accessed in order so that spatial coherence is achieved. The main problems with this approach are that there may be self-shadowing problems due to the warping and the antialiasing may not generate high quality results over an animation.

The shadow depth map generation step may be costly in certain situations, which is why the rest of this section is devoted to discussing strategies to optimize shadow depth map generation. For example, in static scenes of walkthrough animations, a common shortcut renders the shadow depth map just once and reuses it for the entire walkthrough animation. This is because the shadow depth map is generated in world space (typically, except for some of the nonlinear shadow depth map approaches (see Section 2.3.7)), and if the objects and light do not change, the shadow depth map should remain exactly the same for all frames of the animation. Similarly, with moving-object animations, a shortcut involving two shadow depth maps can be generated—one for the static objects (rendered once), and one for the moving objects (reduced complexity but still rendered per frame). During the shadow testing phase (for a point to be shaded), it is necessary to test against both shadow depth maps. This is useful because in some applications, for instance, in several video game contexts, only a few objects in a large environment might be moving, such as people walking through a city of static buildings. As a result, the second shadow depth map tends to be much faster to render as well as to check during shading. However, this does require the knowledge of what is moving or not.

Regardless of the above different flavors for the shadow depth map approach, it is crucial to realize that the shadow depth map generation preprocess employs many, if not all, of the occlusion culling algorithms (several such algorithms are described by Cohen-Or et al. [106]) in order to deal with large data sets very efficiently. These culling algorithms can be used because the shadow depth map generation preprocess is the same as a regular rendering, except that it is from the light source instead of the camera and no shading-related computations are necessary. For example, Govindaraju et al. [201] use a potentially visible set (PVS) algorithm to cull out polygons that are not visible from the camera and are not illuminated. Those culled polygons do not need to be processed from a shadow standpoint any further. A shadow depth map is created to determine if the remaining polygons are fully lit, fully occluded, or partially occluded. Any polygon that is fully lit or occluded is easily handled. Any partially occluded polygon goes through a shadow-polygon clipping process, whereby the polygon is clipped between the lit and shadowed regions. A pixel-based criterion is chosen to avoid

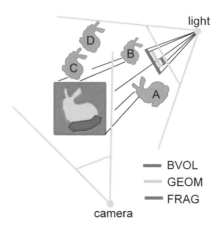

Figure 2.3. Occlusion culling where certain shadow casters can be omitted during shadow depth map generation. ©*2011 ACM, Inc. Included here by permission [52].*

clipping against polygons whose clipped results would not be visible anyway. Due to the above optimizations, the list of such partially occluded polygons should be small, and therefore, the shadow-polygon clipping process should not be too slow. The hybrid approaches can also be split up among three GPUs to render the shadows in real time. However, a numerically robust shadow-polygon clipper remains difficult to implement.

Bittner et al. [52] introduce another shadow-occlusion culling algorithm. First, a rendering from the camera viewpoint is done in order to determine the visible shadow receivers; objects hidden from the camera do not require their shadow status determined. They then compute a mask for each shadow receiver to determine the regions that need shadow computations, thus culling out many shadow casters. The mask can be a bounding volume (BVOL), a rasterization of the object (GEOM), or visible regions of the rasterization (FRAG). In Figure 2.3, the colored bunny is the only visible receiver and the red-highlighted section is the visible region of that bunny. Using the BVOL mask, the bounding box of the visible receiver narrows the angle of view of the light such that D is entirely culled from shadow considerations. Using the GEOM mask, where the rasterization of the visible receiver results in an even smaller angle of view, C and D are culled from shadow considerations. Using the FRAG mask, where the rasterization of only the visible region results in the smallest angle of view, B, C, and D are culled from shadow considerations, leaving only A for processing. In fact, in manually analyzing this scene, one will notice that only parts of A (and the receiver, the colored bunny) need to be processed in the shadow depth map.

Note that when dealing with atmospheric shadows (Section 5.6), many of the occlusion culling techniques discussed in this section may not easily apply.

Temporal Coherence Considerations

To achieve shadows between view frames, Chen and Williams [88] interpolate the results between a fixed set of shadow depth maps so that not all frames require shadow depth map generation. Such a technique can be extended for stereo viewing as well, although the interpolated results tend to be unpredictable and may produce artifacts.

Scherzer et al. [505] use temporal reprojection with a history buffer of shadow depth maps to increase the confidence of the antialiased result and smooth out the transition between frame history to reduce aliasing artifacts of the shadowing. This sounds good in theory, but a number of problems prevent such algorithms from being practical. Among the potential difficulties, one can consider the following: moving objects may not be of much use to the history buffer, still frames can change in appearance due to the different history buffers, objects not in the previous frame cannot contribute to the history buffer and can negatively affect the current frames by promoting the assumption that the first set of frames provide essentially correct results (i.e., bad results will continue to taint the history buffer), and the extra memory needed is significant when using the history buffer, especially for more than a few lights.

Another concern with any algorithm that takes advantage of temporal coherence is its inability to simply send frames out to a render farm in an offline rendering situation because the sequence of frames need to be sequential.

Memory Issues

From a memory standpoint, the storage expense of shadow depth maps for many shadow-casting lights can become prohibitive. One simple way to reduce memory usage is to tile each shadow depth map and compress each tiled shadow depth map on disk. A least-recently-used (LRU) array of tiles is stored in memory, and when a shadow sample is needed, its corresponding tile will be loaded from disk, decompressed, and added to this LRU array. The decompressed tile will typically be useful for a number of consecutive samples. When the LRU array is full, least-used tiles are dumped in favor of incoming tiles. This scheme is effective because the LRU array keeps the memory usage constant, and the decompression of tiled shadow regions is fast. The above generic LRU-tile approach has been implemented in Maya and mentioned in several papers [172, 480, 130]. Note that Ritschel et al. [480] achieve high compression from coherence between multiple shadow depth maps with similar depth values, and Diktas and Sahiner [130] remind us that shadow maps storing surface IDs (Section 2.3.4) can be very efficient to compress and thus can be used in the above LRU scheme as well.

In the PolyTrans software, adaptive tiles are used to reduce memory where only bounding regions of occupied z-regions are stored (instead of the entire shadow depth map) [333]. In Figure 2.4, the objects are actually lying on a floor. The

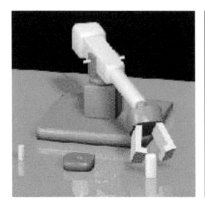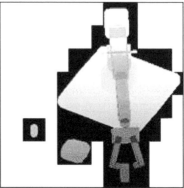

Figure 2.4. Shadow depth map using adaptive tiles to reduce memory usage. *Image courtesy of Okino Computer Graphics.*

floor is not part of the tiling because PolyTrans stores mid-distance (Section 2.3.4) z-values, and therefore the floor becomes exempt from the shadow depth map.

A geometry with different levels of detail (LOD) is a common solution to reducing the memory used in a renderer. However, when using different LOD representations between the camera view and the shadow depth map generation process, there may be bad self-shadowing problems because the depth values of the different representations may cause discrepancies during comparison of depth values. An example of LOD can come from optimized tessellation of higher-order surfaces (such as NURBS or subdivision surfaces) or on-the-fly tessellation of displacement mapped surfaces [110] based on the camera view. In such situations, the pregenerated shadow depth map needs to verify that its optimized tessellation for the light view is not too different from the camera view. Although this problem is an issue for most shadowing algorithms, care must be taken to avoid this problem in particular for the shadow depth map because the shadow depth map generation pass (from the light's viewpoint) is entirely separate from the camera-view rendering.

Another example of different LOD comes from the hybrid rendering of points and polygons. The premise is that for all polygons that are smaller than the area of a pixel they project into, there is no need to spend all the time rasterizing and interpolating polygonal information. In those cases, Chen and Nguyen [87] replace the tiny polygons with points and get much faster rendering times. Shadows were not considered in the paper. Since the choice of point or polygon is determined moment by moment based on the viewpoint, care must be taken to make sure that there is no improper self-shadowing from the tiny polygon that shadows the replaced point. Fixing a depth for each tiny polygon, encoded in the substituted point, may help resolve self-shadowing issues but may also miss proper self-shadowing within the region represented by a single point. Guthe et al. [215] employ perspective shadow depth maps (Section 2.3.7) for the above hybrid rendering but have not addressed self-shadowing issues.

2.3.3 Quality Considerations

There are a number of quality issues related to the standard shadow depth map technique. Figure 2.5 illustrates some of them. These issues will be discussed in the upcoming sections.

- *Avoiding bad self-shadowing problems.* The sphere and cone should be shadow-free. The narrow dark strips (highlighted by the red circles) are due to bad self-shadowing. See Section 2.3.4.

- *Filtering considerations.* The aliased shadows of the sphere and cone on the floor (highlighted by the green circles) can be improved upon with filtering techniques. See Section 2.3.5.

- *Combining self-shadowing and filtering.* The quality of the filtering can be improved by algorithms that combine both filtering and bad self-shadowing mitigation at the same time. See Section 2.3.6.

- *Focusing the resolution.* No matter how well the filtering technique works, it still does not really help when the artifacts are very blocky (highlighted by the blue circles). The problem is due to insufficient resolution, and some techniques address how to permit sufficient resolution in desired locations. See Section 2.3.7.

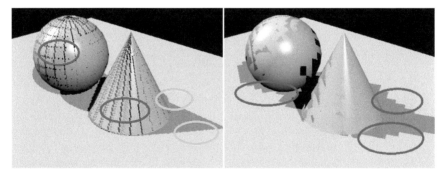

Figure 2.5. Quality issues with shadow depth maps.

Figure 2.6. Moiré artifacts due to bad self-shadowing in shadow depth maps. *Image courtesy of Bobby Anguelov (http://takinginitiative.net).*

2.3.4 Avoiding Bad Self-Shadowing Problems

One rendering quality problem with the basic shadow depth map approach is the bad self-shadowing artifacts that can result from a shadow depth map pixel projecting onto several screen pixels. The resulting artifacts usually appear as aliasing artifacts, either as dark strips on the objects as seen in Figure 2.5 (highlighted by the red circle) or as moiré patterns, as seen in Figure 2.6. The situation is particularly bad when the light direction is close to parallel to the surface being shaded (see Figure 2.7), where Z_n is the shadow depth map pixel value (taken at the center of the pixel), but the pixel itself projects onto a large region on the nearest polygon. As a result, for a point P to be shaded and a point light source L, the test $\|P - L\| > Z_n$ indicates that P is in shadow, but clearly it is not the case. By increasing the shadow depth map resolution, this problem is more localized, but never fully eliminated

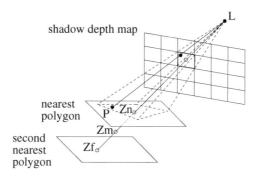

Figure 2.7. Self-shadowing in the shadow depth map algorithm. The distance of point P to the point light L is larger than the depth sampled in the shadow depth map pixel, i.e., $\|P - L\| > Z_n$.

since the situation is exactly the same. Partial solutions to this problem, without a need to increase the shadow depth map resolution, are discussed below, including surface ID, bias, and alternative Z-depth value.

Object or Surface ID

Hourcade and Nicolas [250] address this problem by using a surface ID. Each surface is assigned a unique ID, and each shadow map pixel stores the surface ID that represents the closest surface at that pixel. During rendering, surface IDs are compared instead of Z-depths—if the shadow map pixel has a different surface ID than the current surface, then it is in shadow; otherwise, it is not. However, this general technique does not allow for self-shadowing of surfaces, and thus surfaces must be convex and non-intersecting.

Dietrich [129] and Vlachos et al. [606] apply the self-shadowing surface ID approach on the GPU: surfaces are split up into convex parts and given a unique surface ID, put in a sorted order from front to back with respect to the light source, and the alpha channel is used to compare objects to the corresponding value in the shadow map. However, the sorting and the operation of splitting the surface into convex parts can be expensive and thus is not appropriate for surfaces that morph over time. Alternative hybrid approaches use the surface ID approach for occlusion between different surfaces, and localized shadow depth map or localized shadow volumes are used to determine proper self-shadowing [172].

Bias

Reeves et al. [476] address the self-shadowing artifact using a user-specified offset value called a *bias*. The bias value is used to escape self-shadowing by displacing the closest surface Z_n further away from the light source by a small distance. Thus, there is shadowing only if $\|P - L\| > Z_n + \text{bias}$. However, choosing a good bias value is tricky in many cases because a single bias value must pertain to the entire scene, i.e., a bias value too large can result in shadowed cases where the bias value addition causes the algorithm to indicate fully lit; a bias value too small does not remove the bad self-shadowing. To improve this situation, Reeves et al. [476] introduce a bias value that has a lower and upper bound on biasing, and apply a stochastic factor to have different per-pixel bias value in between, i.e.,

$$\text{bias} = \text{minBias} + \text{random} \times (\text{maxBias} - \text{minBias}).$$

In this way, it is hoped that a good number of cases will properly detect self-shadowing, and for those cases that are inaccurate, it is hoped that the stochastic component will result in noise, which is more acceptable than the aliasing artifacts present in Figure 2.5 (highlighted by the red circle) and Figure 2.6. However, good guesses at bias values are still needed.

Everitt et al. [166] apply a small texture transform offset on the GPU to achieve an offset behavior but cannot achieve a random factor for the offset. In OpenGL (glPolygonOffset) and DirectX (D3DRS_DEPTHBIAS, D3DRS_SLOPESCALE DEPTHBIAS), a slope-scale bias factor has been implemented. This can best be described as

$$\text{bias} = \text{minBias} + m \times \text{slopeScaleBias},$$

where m is another user input that acts as a multiplier to a computed slopeScaleBias value. There are variations of how slopeScaleBias can be calculated, but the general idea is to calculate the surface slope of the polygon as seen from the light; slopeScaleBias needs a larger offset as the slope is closer to the light direction. Wang and Molnar [616] use the tangent plane (i.e., the plane perpendicular to the surface normal) as the basis for the slope; Schuler [509, 510] calculates the slope by evaluating the depth information from the pixel's neighbors; some authors [15, 117] reconstruct the slope from the triangle itself, but this requires storage of additional triangle information within the shadow depth map.

Alternative Z-depth Value

Woo [642] indicates that keeping the closest Z_n value in the shadow depth map may not be the best choice. A Z-depth value between the closest (Z_n) and second closest surface (Z_f) would be a better choice: in practice, the mid-distance Z_m between the two Z-depth values is chosen for implementation (i.e., $Z_m = (Z_n + Z_f)/2$, see Figure 2.7) thus we refer to this approach as the *mid-distance* approach in this book. If there is no second-closest surface, then Z_m is set to some large value. Thus, the check $\|P - L\| > Z_m$ indicates shadowing. In this case, the closest surface would always be in light, and the second closest surface would always be in shadow. This reduces the need to have a good bias value. However, when Z_n and Z_f happen to be very close together, then the original self-shadowing problem may still persist. Thus, it may still be useful to have the shadowing check as $\|P - L\| > Z_m + \text{minBias}$, where minBias is some small value (but less requisite upon in most cases). Even with these modifications, there can be some light-leaking problems at the intersections of surfaces, and at the boundary between surfaces. For example,

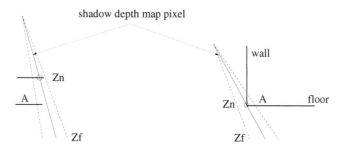

Figure 2.8. Two cases where the mid-distance approach [642] is incorrectly shadowed.

refer to Figure 2.8, where there are two cases where polygon A, under the depth map pixel, is computed as "not in shadow" but is clearly in shadow.

To resolve problems found with the mid-distance extension, Weiskopf and Ertl [627] compute a depth value using a dynamically generated bias value that is the smaller of a (user input) bias value and Z_m. In other words, the shadowing check is $\|P - L\| > \min(Z_m, Z_n + minBias)$. This formulation should avoid the problems seen in Figure 2.8. This approach also requires more storage for both Z_m and Z_n per shadow depth map pixel.

Note that the preprocess render time to achieve mid-distance computations is a bit more complicated and tends to be slower. As a result, Everitt et al. [166] implement the mid-distance variation on the GPU by applying depth peeling [167]. Depth peeling works by first doing the shadow depth map pass, then the nearest surface is "peeled away" in favor of some combination with what is immediately behind it. The mid-distance variation is enticing for the GPU because it relaxes the need for high precision in general numerical computation and in the depth values, as the GPU textures used for the shadow depth map are typically only 16 or 24 bits per pixel.

Wang and Molnar [616] choose the second-closest surface Z_f as the Z-depth value to store in the shadow depth map pixel in order to escape self-shadowing at Z_n. In this case, the second-closest Z-depth Z_f is not concerned about being in light because the $\hat{N} \cdot \hat{L}$ check has already put it in shadow. To compute the second-closest surface hit, performing a front-face culling render does the trick very efficiently, assuming that the surface normals are properly oriented (i.e., outward facing). This approach can also deal effectively with the terminator problem. However, the paper assumes that the surfaces are closed, which is often not the case in many DCC applications, nor is it the case once the closed surface is cross sectioned and not capped (which has become a common feature in CAD and scientific visualization applications). This technique is also less able to handle the cases at silhouette edges or with thin objects, resulting in light-leaking problems in some cases.

Final Words on Self-Shadowing

The self-shadowing problem remains not entirely solved at this point. Each approach has its merits and limitations. However, some variation using the bias approach remains the safest choice for production needs because when something goes wrong (and it inevitably will), there is an "out" via manual input of an appropriate bias value. If atmospheric shadows are needed (Section 5.6.2), or shadowing from voxels (Section 3.7.5), then the bias approach appears to be the best approach.

Unfortunately, we cannot close the book on self-shadowing either. This is because we have to reevaluate the situation for new techniques in combining self-shadowing and filtering (Section 2.3.6) and warping shadow depth maps (Section 2.3.7).

2.3.5   Filtering Considerations

Because the basic shadow depth map uses the typical Z-buffer to compute Z-depth values, the results tend to be aliased. Reeves et al. [476] use percentage-closer filtering (PCF) with a basic Z-buffer to achieve antialiasing (Hourcade and Nicolas [250] employ a similar filtering scheme). Basically, PCF calculates a fractional occlusion value by doing depth value comparisons (between Z_n and $\|P - L\|$) for the current and neighboring shadow depth map pixels (see Figure 2.9 for a 3 × 3 neighboring region). This remains the most cost-effective method and results in blurred shadows rather than correct antialiased shadows, although the rendered shadows are quite pleasing in most cases, and this approach remains the most successful implementation of the filtering techniques (see Figure 2.10 for an image rendered using this technique).

Using PCF, the size of the filter region is user selectable and a large filter region fakes soft shadows, i.e., the softness of the shadows is not physically correct but an attribute of filtering. A typical region size is either 2 × 2 or 3 × 3, where the shadowing amount is the average of the shadowing hits based on each Z-depth comparison (see Figure 2.9). Implementations of PCF on the GPU are discussed in a few papers [65, 75], and most GPUs today can do a 2 × 2 PCF in a single fetch.

In terms of PCF performance, there is no need to sample all pixels inside the $n \times n$ filter region per point to be shaded, as this can become expensive when n is larger than 3. One way is to stochastically choose (fewer than $n \times n$) points in the filter region. Another way is to rotate a fixed pattern of samples [260]. Both techniques can also help reduce banding. Another optimization is introduced by Uralsky [596], where depth comparison is done for a few initial samples, and if the shadow occlusion decision is the same (either all in shadow or all lit), then no more samples are computed and the same shadow occlusion decision is used.

In terms of quality improvements over the standard PCF, Pagot et al. [450] use multiple depths to better filter the results, but the visual results do not look that promising. Zhao et al. [671] use the alpha coverage values to better filter the results to mitigate the sawtooth artifacts from a typical PCF interpolation. The basic idea

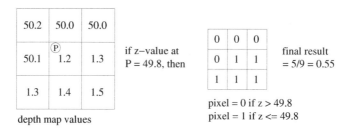

Figure 2.9. Percentage-closer filtering is used to compute fractional occlusions, resulting in good filtered results.

Figure 2.10. Shadow depth map rendering with bias offset and percentage-closer filtering, which appears much cleaner than the equivalent shadows in Figure 2.5(left).

is to use the alpha values to better estimate the occluder silhouette and incorporate it into the PCF equation. However, this works well primarily for a single occluder within the PCF kernel. Kim et al. [302] apply bilateral filtering to achieve better results. Bilateral filtering is basically a nonlinear product of two Gaussian filters in different domains, which has been known to generate smooth images while preserving strong edges. Although the above solutions have improved some of the resulting shadows, prefiltering techniques (Section 2.3.6) have had better adoption so far.

Instead of using PCF, Keating [298] stores fragments per pixel, a depth value per fragment, and employs an A-buffer [82] to antialias shadow edges. Because there is fragment information per shadow depth map pixel, the shadow depth map resolution can be decreased to get a similar quality. Similarly, other buffer variations to achieve antialiased shadows include the cross-scan buffer [576], EE buffer [600], and ED buffer [261]. Note that the above techniques require additional information on top of a standard shadow depth map.

Another way to filter the shadow results comes from using shadow textures [300]. While the shadow depth map generation pass is still required, its results are stored in a UV texture atlas for each object. In this way, the filtering used is the same for texture mapping (e.g., mipmap within the GPU). However, this algorithm requires a lot more memory and processing time to map the shadow depth map generation results into each object's UV texture space.

2.3.6   Combining Self-Shadowing and Filtering

Recently, some interesting research into better quality prefiltering of the shadow depth map has surfaced, including the variance shadow map, convolution shadow

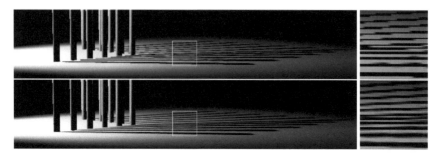

Figure 2.11. PCF (top) versus variance shadow mapping (bottom). ©2006 ACM, Inc. Included here by permission [140].

map, and exponential shadow map. These methods use a probabilistic approach and attempt to solve both the filtering and self-shadowing problems within a single algorithm. They also generate superior-quality results.

With the variance shadow map (VSM) [140], instead of storing a depth value per pixel as in the standard approach, each pixel stores the mean μ and mean squared value of a distribution of depths, from which the variance σ^2 can be derived. The mean is computed during an additional prefiltering step of the surrounding pixels of the shadow depth map—the softness of the shadow is determined by the size of prefiltering. When shading a particular point, the shadow occlusion fraction is computed by applying the one-tailed version of the Chebychev's inequality theorem, which represents a statistical upper bound for the amount of occlusion:

$$\text{unoccludedlight} = \frac{\sigma^2}{\sigma^2 + (\|P - L\| - \mu)^2}.$$

The main advantages of this algorithm are that the filtering and self-shadowing problems are solved in a single approach, only a single shadow map pixel is accessed per shading step (thus it is much faster when the filtering region is large), and the shading step should be more pleasing with thin features than percentage-closer filtering [476]. See Figure 2.11 for a visual example of the quality improvement. However, the disadvantages are that the increase in shadow map memory usage doubles over the standard shadow depth map approach and that the upper bound property used as shadow occlusion often makes the shadows less dark than expected, thus making the objects incorrectly appear to float in midair. They also identify a high frequency, light-leaking problem due to large variance cases from large differences in depth values, which can be manually improved through a user-input *bleeding factor*. See Figure 2.12 for a visual example of the light-leaking problem. Improvements to the light-leaking problem are achieved by slicing the scene into depth intervals with better depth accuracy, using summed-area tables [334], but this comes at the expense of even larger memory requirements. This requires the user to input the number of depth intervals for acceptable results.

Figure 2.12. Light-leaking problem in the original variance shadow map implementation (left). Slicing into depth intervals reduces the light-leaking problems (right). *Image courtesy of Lauritzen and McCool [334].*

Very recently, a variation of the VSM can be seen in the work of Gumbau et al. [211] using a cumulative distribution function (CDF), in which the Chebychev's inequality function is replaced with a Gaussian or power function. Gumbau et al. indicate a reduction in light-leaking problems as compared to VSM but also suggest using a number of depth intervals to resolve these problems.

Convolution shadow mapping (CSM) [11] approximates the shadow occlusion fraction using a 1D Fourier series expansion. Each shadow map pixel is converted into several basis textures instead of a single depth value. During the shading of a point, prefiltered texture samples are used to smooth the shadow result. In other words,

$$\text{unoccludedlight} = \left[w * f(\|P - L\|, Z_n) \right](d),$$

where d represents the shadow map pixel, f the Fourier series expansion, w the filter kernel, and $*$ the convolution operator. As compared to the variance shadow map, the convolution shadow map is slower (due to the spontaneous convolution) and uses much more memory (to store the basis textures). However, it does not exhibit the type of light-leaking problems as seen in VSM, although it does exhibit ringing artifacts, which can be manually improved by user inputs on the series expansion order and an *absorption factor*, respectively.

The above problems led to the proposal of the exponential shadow map (ESM) [13, 496], where an exponential expansion is used instead of the Fourier series expansion—i.e.,

$$f = \exp(-c\|P - L\|)\exp(cZ_n).$$

This shorter expansion allows the memory requirements to be significantly less than the convolution shadow map and also causes faster convolution computations, and no light-leaking problems, although there is guesswork as to what the appropriate value for c might be, which can change the shadow appearance quite

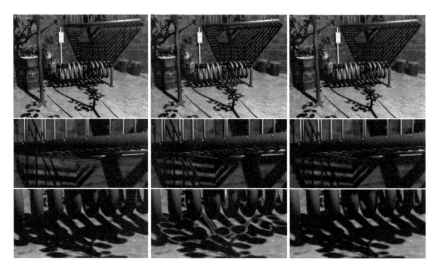

Figure 2.13. Rendering of the same scene using CSM (left), VSM (center), and ESM (right). Main CSM artifacts are near the stairs where shadows fade, and main VSM artifacts are from the light-leaking problems. *Image courtesy of Annen et al. [13].*

drastically. An example comparison of the three techniques can be seen in Figure 2.13.

In terms of comparisons of the three above approaches, Lv et al. [373] evaluate both VSM and ESM and find that computing both approaches, and then taking the minimum of the occlusion values resolves much of the quality limitations of both. However, this is expensive to implement. Bavoil [43] notes that from a performance perspective, the CSM is much slower, whereas the VSM and ESM are about the same in performance. What is very important is that there are at least one or two additional parameters needed for each approach to manually fix visual artifacts, which can be a hinderance during production. However, with better quality and performance results over standard percentage-closer filtering, these techniques are quite worthwhile, and VSM appears to be the most often adopted technique of the three (VSM, CSM, ESM).

One interesting side effect of solving the self-shadowing and filtering in this manner is that the terminator problem (Section 1.4.1) now rears its ugly head because the bias offset is no longer available to escape the self-shadowing (deeper subdivision is the main workaround). VSM and ESM can also exhibit terminator artifacts slightly differently, with sawtooth effects (see Figure 2.14). The unexpected artifact from PCF is the bad self-shadowing at the polygon edges: this is due to the drastically different Z-depth values near the polygon edges (when extremely undertessellated), causing incorrect self-shadowing. This can be alleviated with a larger bias value or one of the alternate-Z self-shadowing correction techniques (Section 2.3.4).

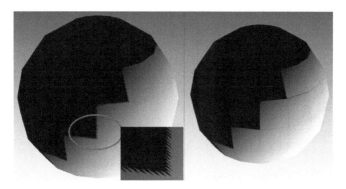

Figure 2.14. Terminator problem shown in VSM/ESM (left) and in PCF (right). *Image courtesy of Thomas Annen.*

2.3.7   Focusing the Resolution

It is prudent to consider rendering the smallest region possible for the preprocess of shadow depth map generation. This gives higher-quality results for a specific resolution. Brabec et al. [69] clip the region for shadow depth mapping to be the intersection of an optimized per-frame view frustum and light frustum. The optimized per-frame frustum results in a smaller angle of view as well as a smaller z-range for clipping (near and far clipping planes). See Figure 2.15 for an example of such a reduction in the angle of view and z-range clipping. Additional reduction may be achieved with the occlusion culling techniques described in Section 2.3.2. However, keep in mind that quality can change for the same shadow depth map resolution if the angle of view region to be rendered changes; thus, this approach should be considered for entire units at a time, not objects that would fly away and back during an animation (thus dramatically changing the size of the region to be rendered). See Figure 2.16, where the same shadow depth map can result in

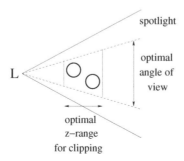

Figure 2.15. Given a spotlight, reduce it to the relevant angle of view and z-range clipping for optimal shadow depth map rendering.

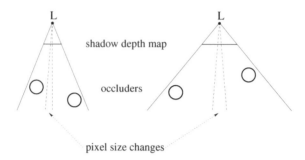

Figure 2.16. In an animated sequence, the size of a pixel in a shadow depth map with the same resolution can be altered due to changes in the optimized angle of view.

different quality due to the different coverage of 3D space; the shadow depth map pixel between the two cases is of different sizes. This can be partially resolved by adjusting the shadow depth map resolution to match the percentage of the smaller angle of view to the original angle of view [599, 668]. Further, dramatic changes in z-range for clipping can produce temporal artifacts, as this changes the depth precision over an animation.

Another approach to focusing the resolution is to remove the usually large floor from the shadow depth map computation and focus the shadow depth map on the much smaller objects above the floor. This assumes that the floor does not self-shadow, nor does the floor shadow other objects. Shadow z-comparisons are still being done to the floor and can capture the shadow information from the objects above it (i.e., the floor can receive shadows). For example, for a car moving on a large, static road, it is not necessary to include the road in the shadow depth map if only the car casts its shadow on the road. As a result, the shadow of the car on the road should be of very high quality.

The above optimizations can still be limiting and may require high-resolution shadow depth maps. Nonlinear shadow depth map alternatives, to be discussed in the next sections, can achieve good quality while maintaining reasonable shadow depth map resolutions. These approaches mitigate the blockiness artifacts in Figure 2.5 (indicated by blue circles).

Nonlinear Shadow Depth Maps

There are currently several categories of nonlinear shadow depth maps, each category focusing the resolution to resolve slightly different issues. They are discussed in the upcoming sections.

- *Warping shadow depth maps.* The shadow depth map is warped such that the focus of the resolution is higher near the viewpoint and lower farther away from the viewpoint. This improves what is referred to as perspective aliasing, where shadow blockiness is mitigated for view-nearby points, but extreme care must be taken to avoid worsening quality for far-away points that exhibit high-frequency shadow changes; thus, this technique may not be appropriate for high-quality offline rendering scenarios.

- *Z-partitioning shadow depth maps.* This is a nonwarping alternative approach to mitigate perspective aliasing where zones along the Z-depth axis are partitioned into separate depth maps, with high resolution focused on the Z-depths closer to the camera view. This approach should produce better quality than warping shadow depth maps, although at a slower performance rate. Note that this approach is known by various names, including sunlight buffers, parallel-split planes, z-partitioning, or cascaded shadow depth mapping.

- *Alias-free shadow maps.* This method calculates shadowing only for the needed visible points. It represents the highest-quality approach to address perspective aliasing, although it cannot easily adopt filtering techniques (discussed in Sections 2.3.5 and 2.3.6) and changes the shadow map rendering pipeline quite significantly. It also provides extensions to soft shadows (Section 4.5.5) and shadows for highly complex thin materials (Section 5.5).

- *Adaptive shadow depth maps.* For this approach, the higher resolution is focused on shadow edges where the insufficient resolution would result in shadow blockiness. Thus, even in the tricky case when the light direction is nearly perpendicular to the normal of the point to be shaded, the increase in focus on shadow edges mitigates shadow blockiness as a result of projection aliasing. These approaches tend to produce higher-quality results than warping and z-partitioning approaches, but the performance is slower than both approaches.

Some of these nonlinear approaches have been used successfully in real-time applications. However, robustness of the approaches should be more carefully examined. For example, the directional light is the main light type discussed when dealing with warping shadow depth maps, and better examination into spot and point lights is warranted. For all the above approaches, it is unknown how a point light, under six shadow depth maps forming a cube, will perform in terms of discontinuity artifacts at the boundaries of the six shadow depth maps. As well, although approaches addressing perspective and projection aliasing can be combined, it is not clear whether it is worth the trouble.

Warping Shadow Depth Maps

Stamminger and Drettakis [557] introduce perspective shadow depth maps (PSM) to concentrate the depth map shadow details according to the current view by computing and comparing depths in postperspective space, i.e., they squash the view into a $[(-1, -1, -1), (1, 1, 1)]$ clipping cube (see Figure 2.17(top)). Thus we can get

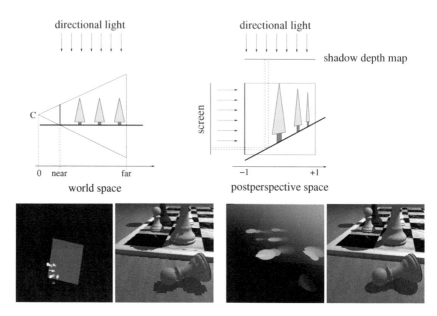

Figure 2.17. In the top row, schematic view world space and postperspective space. In the bottom row, standard shadow depth map (left) versus perspective shadow depth map (right). ©2002 ACM, Inc. Included here by permission [557].

very high precision and high quality with respect to shadows seen close to the camera view. The shadows far away from the camera view are coarser, but then their coverage in screen space is small, so it should not matter as much. However, to get high precision for objects close to the camera, it is critical to have a tight near clipping plane so that the concentration of precision is not wasted on empty space ahead of the near clipping plane. This approach maps well to the GPU but cannot render the shadow depth map once for walkthrough animations (as discussed in Section 2.3.2).

See Figure 2.17(bottom), in which the two leftmost images represent the regular shadow depth maps, whereas the perspective and current view nature (as seen in the two rightmost images) allow the shadow depth map to concentrate on a certain region and to cover more details in that region.

Perspective shadow depth maps have since attracted much attention. However, they have a number of quality issues, among them,

1. Known as the virtual camera issue, i.e., shadow occluders located behind the camera are not numerically stable in the perspective space. Stamminger and Drettakis [557] suggest shifting the camera back to handle such occluders. However, the numerical behavior will worsen and produce poor-quality results.

2. The shadow quality depends heavily on the light direction relative to the camera view [317]. If the light direction points towards the camera, the shadow image quality can be poor. Chong and Gortler [95] also indicate poor quality in common situations when the object is far from the camera.

3. It is not immediately clear how warped the shadow depth map parameterization should be for the best overall quality.

4. Because of the need (on a per-frame basis) to adjust the near clipping plane to optimize precision and quality, Martin [387] indicates that changes in the near and far clipping planes over an animated sequence may also cause shifts in quality.

5. Martin [387] and Kozlov [317] indicate that due to the nonlinear behavior of the Z-depth values, it is very difficult to come up with a good bias value. The topic of self-shadowing thus needs to be revisited beyond what has already been discussed in Section 2.3.4 for regular shadow depth maps.

Kozlov [317] focuses on the special cases that make perspective shadow depth maps difficult to robustly implement. For the virtual camera issue (issue 1), the postperspective camera transformation is reformulated such that a slightly larger area of coverage is achieved without having to extend back the light frustum and cause as much image quality sacrifice. Chong and Gortler [97, 95, 96] interpret the perspective shadow depth map as an orientation selection of the shadow depth map projection plane, according to the visible points in the current view. They present metrics to automatically choose an optimal orientation in 2D, indicating that these metrics could be extended in normal 3D environments.

To improve on issues 1 and 2, trapezoidal shadow depth map (TSM) [386] and light space perspective shadow depth map (LiSPSM) [633] apply a projection for the light source and then a projection along the camera view, projected into the view plane of the light. Thus, the virtual camera C becomes perpendicular to the light's view direction. See Figure 2.18, where C is created on the left, and the resulting warp on the right.

For warping in issue 3, LiSPSM [633, 632] provides a user-controlled parameter n that pushes the viewpoint back. A small n value indicates a stronger warp and resolution focus on nearby objects. Wimmer et al. [633] also attempt to employ a logarithmic perspective transformation because of the ideal warping, but the rendering of logarithmic shadow depth maps proves to be expensive and difficult to support on the GPU. Other authors continue to investigate variations and enhance the performance of the logarithmic perspective transformation [665, 359, 360], but none are feasible on the GPU yet.

For TSM [386], the light frustum is approximated as a trapezoid, and shadow depth values are represented in trapezoidal space. Relating to issue 4, the danger of concentrating shadow depth map information resides in the fact that there can be

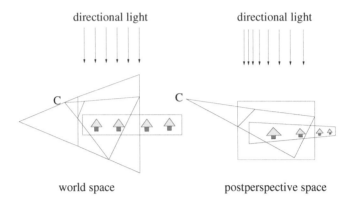

Figure 2.18. LiSPSM fundamentals, with the concept of the virtual camera C.

jumps in quality over a sequence of frames because the region of the concentration can change significantly over a frame. A scheme is devised to lower the frequency of the jumps by having the trapezoidal transformation map the focus region to within 80% of the shadow depth map (see Figure 2.19). This heuristic decreases the number of jumps but does not eliminate them.

As for the self-shadowing issue (issue 5), the artifacts appear most pronounced for TSM, and Martin and Tan [386] suggest linear z-distribution by omitting the z-coordinate from the perspective transform. Kozlov [317] indicates that a single bias value is unable to resolve the self-shadowing issue well, and thus proposes a slope-scale bias instead. Gusev [214] reformulates the shadow depth map parameterization and refers to this as extended perspective shadow depth map (XPSM). Based on this parameterization, the bias calculation can be simplified with a single bias value. Scherzer [508] experiments with various Z-depth value approaches

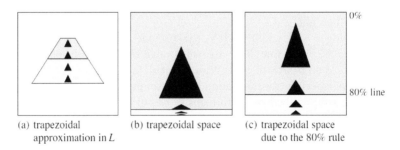

Figure 2.19. Trapezoidal shadow depth map focuses a region to reduce jumps over an animation. *Image courtesy of Martin and Tan [386], ©Eurographics Association 2004. Reproduced by permission of the Eurographics Association.*

(Section 2.3.4), and finds that the best results come from using the combination of linear z-distribution and slope-scale bias.

Lloyd et al. [358] compare the error distribution and results of the PSM, LiSPSM, and TSM, and found the same overall error. The LiSPSM approach generates the most even distribution of errors and thus has been the favored area of continued research among the warping shadow depth map approaches. A thorough mathematical treatment of the warping shadow depth maps is provided by Eisemann et al. [158].

In general, the above issues show improvement, but are not completely solved. Furthermore, robust handling of near clipping-plane changes, camera looking at the light source, and discontinuity issues surrounding omnidirectional lights have not been pursued. Finally, the sacrifice of lower resolutions at farther away distances can also cause noticeable aliasing artifacts that may not be acceptable for offline rendering animations.

Z-Partitioning Shadow Depth Maps

Before we begin this section, we want to remind the reader that the techniques described in this section go by multiple names, including sunlight buffers, parallel-split planes, z-partitioning, and cascaded shadow depth map.

Several authors [575, 663, 664, 358, 161, 131] propose a set of parallel-split (to the view plane) planes, where a cluster of objects is tightly bound by the parallel-split planes and a separate shadow depth map is generated per parallel-split plane. The resolution of each such shadow depth map is dependent on the distance from the viewpoint: the closer the parallel-split plane is to the viewpoint, the higher the resolution, thus capturing the near details of interest with high resolution. See Figure 2.20.

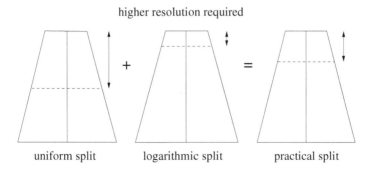

Figure 2.20. Splitting the shadow depth map regions into parallel-split planes, using (from left to right) uniform subdivision, logarithmic subdivision [358], and weighted average of the two types of subdivision [664].

The main questions are how many and where to position the parallel-split planes:

- Valient [598] recommends some manual intervention.

- Lloyd et al. [358] indicate that a logarithmic scale subdivision is ideal due to the logarithmic perspective transform. See Figure 2.20(middle).

- While a logarithmic subdivision may be ideal, it is not appropriate because most of the resolution would be close to the near-plane; thus, Zhang et al. [664] suggest some weighted average between the above logarithmic approach and an equidistant split-plane distribution. See Figure 2.20(right).

- Lauritzen et al. [335] indicate that a static determination of the subdivision locations can either be wasteful or ineffective and suggest a scene-dependent logarithmic approach, where the optimal subdivisions are computed based on evaluation of the actual light-space shadow depth map sample distribution.

- Independent authors [35, 408] use two parallel-split planes in conjunction with a light-space perspective shadow depth map, so the accuracy of the split is not as critical.

As compared to warping shadow depth maps, the methods of parallel-split planes appear to have fewer special cases that they need to deal with, such as light being behind the viewpoint, resulting in better quality rendering. However, care must be taken to reduce visual discontinuity artifacts at the borders of the parallel-split planes by blending between the bordering parallel-split planes [668].

From a performance standpoint, the technique should be slower than the approach of warping shadow depth maps, because it takes longer for shadow depth map generation (improvements are discussed by Zhang et al. [667]), as well as shading of a point (multiple shadow depth maps may need to be accessed to compute shadowing).

Alias-Free Shadow Maps

Aila and Laine [4] (who refer to their approach as alias-free shadow maps), and Johnson et al. [275] (who refer to their approach as irregular shadow mapping) independently come up with an approach where the actual shadow depth map is not required any more.

An initial rendering from the camera is done—this step identifies the points P to be shaded, in which shadow computations are needed. All points P are transformed to light space (P'), and their corresponding Z_n (near Z-depth value) is computed, of which Z_n is then compared against the distance $\|P' - L\|$ to determine shadowing occlusion. The obvious solution to compute all Z_n is to use ray

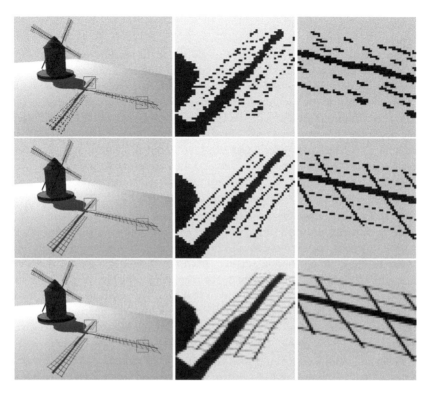

Figure 2.21. Visual comparison of the conventional shadow depth map (top), alias-free shadow map (middle), and subpixel alias-free shadow map (bottom). *Image courtesy of Pan et al. [451]. Computer Graphics Forum ©2009 The Eurographics Association and Blackwell Publishing Ltd. Published by Blackwell Publishing, 9600 Garsington Road, Oxford OX4 2DQ, UK and 350 Main Street, Malden, MA 02148, USA. Reproduced by permission of the Eurographics Association and Blackwell Publishing Ltd.*

tracing, but that is too slow. Instead, Johnson et al. [275] use a nonuniform Z-buffer for rasterization, and Aila and Laine [4] use a 2D BSP tree.

While this approach is promising, the different preprocess step is slower than the standard shadow depth mapping, and nontrivial effort is needed to apply some simple percentage-closer filtering [476]. To mitigate the slower performance, some authors [21, 535] detail a GPU implementation of the above approach. Pan et al. [451] also build on top of the alias-free shadow maps by considering the subpixel-level aliasing that is produced by the rasterization on the view plane. To retain good performance, a silhouette mask map is used to do only subpixel computations on silhouettes. See Figure 2.21 for a comparison of the conventional shadow depth map, alias-free shadow map, and the subpixel alias-free shadow map.

Adaptive Shadow Depth Maps

Fernando et al. [172] apply an adaptive resolution to the shadow depth map, focusing more resolution on the shadow boundaries using a quadtree structure. This way, they do not need to store large shadow depth maps to get high quality. The shadow boundaries are identified using edge-finding techniques. The criteria for higher resolution subdivision is Z-depth discontinuity, provided the shadow depth map resolution is lower than the camera-view resolution. Lefohn et al. [341, 342] avoid the slow edge-finding techniques by identifying all shadow depth map pixels that are needed from screen space pixels and then sample at a higher resolution for those shadow depth map pixels. Coherence between camera-space, light-space, and GPU usage are also discussed to optimize the algorithm.

Arvo [20] uses tiled shadow depth maps, where the light view is initially divided up into equally sized tiles. The observation is that each tile should not be considered equal, i.e., each tile should be computed with a different resolution to maximize the eventual shadow quality. Each tile computes a weight that determines the resolution of the shadow depth map, where the tile weight is the summation of the weight of each pixel (within that tile). The pixel weight is a function of the depth discontinuity (if it contains a shadow boundary), depth difference (between the occluder and receiver), and surface distance (between the surface and camera view—higher resolution when the surface is close to the camera view).

Giegl and Wimmer [197] introduce queried virtual shadow depth maps, where an initial shadow depth map is generated, then tiled, and where each tile is shadow-rendered at double the resolution, then compared to the previous shadow-render. If the differences in shadow depth map pixels between the shadow-render passes are large, then another shadow-render is done at twice the resolution. This is recursively repeated until the difference is below a certain threshold. Giegl and Wimmer [196] introduce fitted virtual shadow depth maps to reduce the large number of shadow-render passes needed for the queried virtual shadow depth maps by trying to predetermine the final refinement levels for the quadtree; this is done by prerendering the (shadow) scene and computing the necessary resolution needed for each 32×32 tile based on the Z-depth differences for the screen locations. See Figure 2.22 for a rendering of the approach compared to standard shadow depth maps at very high resolution (4096×4096).

On a slightly different track for adaptive depth maps, Sen et al. [518] generate a piecewise linear approximation of the occluder's silhouette region, so that the advantages of shadow depth map are retained, while the silhouette approximation results in high-quality shadow edges. A standard shadow depth map is created as usual, along with a silhouette map, which is computed using a dual contouring algorithm to store shadow-edge information. During shading, if the four closest samples all agree on the shadowing occlusion (i.e., all in shadow or all in light), then this pixel is considered in shadow or in light without any further computations. However, if there are shadow occlusion differences, then the silhouette is found,

Figure 2.22. Standard shadow depth map at a resolution of 4096 × 4096 (top) versus fitted virtual shadow depth map (bottom). *Image courtesy of Giegl and Wimmer [196]*.

and the silhouette map is used to determine the exact amount of shadowing for the current pixel. The main issue with this approach is that when edges are too close to each other, boundary artifacts can remain because the shadow edges may not be easily identifiable. Similarly, Bujnak [74] also focuses just on the silhouette to achieve better shadow-edge quality. Instead of shadow edges as in Sen et al. [518], silhouette information is encoded inside the shadow depth map using one or two quads per silhouette edge.

2.4   Shadow Volumes

Although the earlier description of shadow volumes in Section 1.3.3 appears simple, details as to how to compute the initial shadow count, how to intersect the shadow polygons to update the shadow count, etc., require elaboration for robust implementation. In fact, such implementations are separately covered for the CPU (Section 2.4.1) and GPU (Section 2.4.2) because they need to be handled quite differently. We then discuss performance-related algorithms to reduce the shadow polygon complexity and fill rate, which are applicable to both CPU and

GPU (Section 2.4.3). The shadow volume discussion concludes with the description of shadow volume algorithms that integrate BSP trees (Section 2.4.4).

2.4.1   Shadow Volumes on the CPU

The Initial Shadow Count

When the camera is outside all shadow volumes, then the initial shadow count is 0. Otherwise, the initial shadow count needs to be correctly set as the number of shadow volumes that contain the viewing position. In addition to generalizing shadow volumes for nonplanar polygons, Bergeron [46] indicates the need to close (cap) shadow volumes so that a correct initial shadow count of shadow volumes within the camera can be computed. By rendering a single (arbitrary) pixel and counting all shadow polygons that pierce this ray (even the shadow polygons occluded by the scene geometry), call this E_{svc}, then the initial shadow count is simply $(-E_{svc})$.

Another simpler way is to treat the shadow count calculation as a ray tracing problem in which a ray is shot toward the light from the viewpoint in order to count the number of hits E_{svc} that the ray encounters on surfaces that are front-facing to the light source. In this case, E_{svc} is the initial shadow count. Since this only needs to be done once per rendering, this expensive ray tracing solution can be considered quite reasonable.

For software-based solutions, the initial shadow count can be a single value. In GPU variations, where the shadow polygons are clipped to the near clipping plane, the initial shadow count may be different for particular pixels (see Section 2.4.2 for further coverage on how to deal with initial shadow counts when using GPU).

Visibility Algorithms to Intersect Shadow Polygons

The easiest solution is to employ a Z-buffer for shadow-count incrementing and decrementing [184, 181]. The Z-buffer is first used to render the non-shadowed pass. The shadow polygons are then scan-converted onto the Z-buffer, and any shadow polygon that resides in front of the Z-depth is used to increment or decrement the shadow count. A variation of the Z-buffer approach to employ a scanline-based solution has also been achieved [46].

Slater [537] suggests that it might also be interesting to apply a light buffer [219] approach (Section 2.5.1), in which a hemicube surrounds the light source. Each cell (pixel) in this light buffer indicates the objects that project in it. Thus, to determine the shadow count, the camera ray is projected to the light buffer, and only the relevant cells are visited to access a small candidate set of shadow polygons that the camera ray might hit. Slater [537] compares this light buffer approach with the SVBSP approach [91] (explained in Section 2.4.4) and finds no conclusive evidence about which approach provides faster rendering speeds.

Eo and Kyung [164] suggest a ray tracing solution in which the shadow polygons are scan-converted (distributed) into a voxel data structure and the viewing ray traverses the voxels to find the candidate shadow polygons it should intersect to increment or decrement the shadow count. Of course, this is quite slow, as Eo and Kyung are trying to use it in conjunction with regular voxel traversal (see Section 2.5.1).

The Z-buffer and, in some cases, scanline-based variations tend to be the more commonly accepted implementations. In Section 2.4.2, we show that for shadow volumes on the GPU, the Z-buffer is employed almost universally in the form of a stencil buffer. Due to the Z-buffer and stencil buffer approaches, the only ways to antialias the shadow results is to increase the resolution of the required buffers. The exception is the slower ray tracing approach [164], in which adaptive and/or stochastic samples can be used to antialias shadows.

2.4.2   Shadow Volumes on the GPU

Z-pass Algorithms

Heidmann [236] is one of the first to present a GPU implementation of the original Crow [114] approach. This is referred as the z-pass approach. The exact steps to achieve z-pass shadow volumes include

1. Enable the depth buffer in order to render the scene (without shadows) onto the color buffer.

2. Clear the stencil buffer and disable depth buffering.

3. Draw all front-facing shadow polygons, and increment the shadow count for the relevant pixels in the stencil buffer.

4. Draw all back-facing shadow polygons and decrement the shadow count for the relevant pixels in the stencil buffer.

5. Darken the pixels in the color buffer where the stencil values are not equal to 0.

The above steps, however, do not properly handle the combination of ambient, diffuse, and specular terms. An alternate would be to change steps 1 and 5:

1. Enable the depth buffer in order to render the scene without shadows but only with the ambient term onto the color buffer.

5. Draw the diffuse and specular terms for the pixels when the stencil value is equal to 0.

Unfortunately when using the GPU, shadow polygons must be clipped against the view frustum, thus potentially introducing erroneous initial shadow counts for some of the image pixels if the camera view is inside the scene. One simple way to resolve this is to compute the value E_{svc} for the camera point in terms of

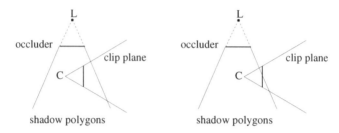

Figure 2.23. Potential errors in the initialization of a single camera shadow count when the near clipping plane intersects some shadow polygons.

its initial shadow count, then initialize the stencil buffer with E_{svc} instead of 0. This method is fairly accurate unless the camera is deep inside the scene and the near clipping plane intersects a shadow polygon. For example, Figure 2.23(left) shows that a single initial shadow count is valid, but this is not the case for Figure 2.23(right), where the shadow polygon cuts across the near clipping plane and the initial shadow counts are different when the near clipping plane intersects the shadow polygon. See Figure 2.24(left) for shadow errors when a single shadow count is assumed, and Figure 2.24(right) illustrates the correct shadow result.

Some authors [37, 393, 48] propose to cap the shadow volumes with new shadow polygons where the shadow volumes are clipped against the view frustum's near clipping plane. Everitt and Kilgard [165] explain why it is difficult to robustly implement this capping technique that needs to account for all the different combinations. However, ZP+ shadow volumes [249] manage to properly initialize all the shadow counts in the stencil buffer by rendering the scene that resides between the light and the camera near clip plane from the perspective of the light. Numerical precision issues can occur when the light or an object is very close to the camera near clipping plane or an object. Z-fail algorithms and per-triangle shadow volumes (both described in the remainder of this section) seem more appropriate for robust behaviors.

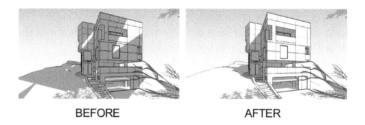

BEFORE   AFTER

Figure 2.24. Shadow errors due to a single shadow count (left), with the correct shadows (right). *Image posted by Aidan Chopra, courtesy of Google Sketchup and Phil Rader.*

Z-fail Algorithms

Another set of solutions [50, 81, 165] to deal with a correct initial shadow count starts from the observation that this point-in-shadow-volume test need not be evaluated along the line of sight, but can also be evaluated from the visible surface to infinity instead (called *z-fail test*). Thus, the shadow count for the viewpoint is simply initialized to 0: occluded back-facing shadow polygons increment their shadow counts, occluded front-facing ones decrement them, and shadow polygons need not be clipped by the near plane. A shadow count greater than 0 indicates shadowing. As can be seen in Figure 2.25(right), the z-fail count starts from the right at 0; then, going left, it ends up at 2 when it hits the point P to be shaded. Note that z-fail can be considered the computation of the z-pass in the reverse direction, where the z-pass (Figure 2.25(left)) starts at the left (camera) at 1, going right, and ending up at 2 when it reaches P.

Note that capping shadow volumes against the rest of the view frustum is still necessary to produce correct shadow counts. Some hardware extensions [301, 165] avoid the need to correctly compute the capping with respect to the view frustum. This has actually become standard in both DirectX (DepthClipEnable) and OpenGL (DepthClamp). Another solution simply eliminates the far clipping process with a small penalty to depth precision; as a bonus, capping at infinity for infinite light sources is not necessary, as all shadow polygons converge to the same vanishing point.

Everitt and Kilgard [165] and Lengyel [345] realize that the z-pass approach is generally much faster than the z-fail approach. They thus detect whether the camera is within a shadow volume and employ z-pass if not, and z-fail if so. However, the sudden shifts in speed difference may be disruptive to some applications. Similarly, Laine [330] determines such cases on a per-tile basis so that the z-pass approach can be used more often than the z-fail approach, while retaining correct shadows. This is done by comparing the contents of a low-resolution shadow depth map against an automatically constructed split plane.

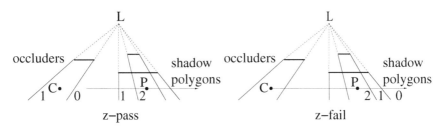

Figure 2.25. Z-pass (left) versus z-fail (right) algorithm. Both algorithms arrive at the same shadow count of 2, but come from opposite directions.

Per-triangle Shadow Volumes

Very recently, Sintorn et al. [536] revisit the shadow volume algorithm, where each scene triangle generates its shadow volume for a point light source. The main idea is to compute a hierarchical Z-buffer of the image. In other words, a hierarchy of min-max 3D boxes is constructed from hierarchical rectangular regions of the image. Pixels with their shadowing status already known (e.g., background, back-facing the light, etc.) are not part of a 3D box. Shadowing is applied as deferred shading.

All four support planes of a triangle shadow volume are rasterized together, and each min-max 3D box is tested against these four planes. Usually, a box will be efficiently culled as not intersecting the triangle shadow volume. If a box is completely enclosed in the shadow volume, it is marked (and all its pixels) as in shadow. Otherwise, the box is refined at the next (more detailed) level of boxes.

The rasterization of the triangle shadow volumes is made very efficient and is computed in the homogeneous clip space, thus avoiding problems with near and far clipping planes. Because each triangle shadow volume is treated independently, the method achieves stable frame rates, and it can work with any soup of polygons. Therefore, it can handle some ill-formed geometry and does not need to prepro-cess silhouette edges. It can be extended for textured and semitransparent shadow casters.

The method, implemented in CUDA, is always more competitive than z-fail methods, and for larger resolution images and/or higher antialiasing sampling, it outperforms z-pass methods. Although it could be less efficient for scenes with high depth variance and tiny triangles, the robustness of the method, its small memory footprint, and its generality show great promise. It is also not immediately obvious if it is worthwhile to consider the optimizations mentioned in Section 2.4.3 for this approach.

Other Considerations

Brabec and Seidel [67] use the GPU to compute the actual shadow polygons to further speed up shadow computations as well as to avoid differences in numerical properties (between the CPU and GPU) that can result in some shadow-leaking problems. In fact, numerical instability issues can potentially arise due to shadow polygons created from elongated polygons, or when the normal of the polygon is almost perpendicular to the light-source direction. Care must be taken in dealing with such cases.

If the silhouette optimization techniques discussed in Section 2.4.3 are not used, then there are likely other numerical problems. In particular, because all shadow polygons (not just on the silhouettes) are created for shadow volumes, there are likely visibility problems due to the limited precision of the z-values—this problem is sometimes referred to as z-fighting.

Roettger et al. [487] replace the stencil buffer with the alpha or screen buffers (because the stencil buffer can be overloaded in its use to cause bottlenecks) and the computations of shadow counts by blending operations, i.e., instead of incrementing and decrementing shadow counts, they multiply and divide by 2 in the alpha buffer, respectively. As the shadow-polygon fill rate is the main bottleneck in shadow volume approaches, these two buffers can be computed at lower resolutions (at the additional cost of a rerendering at this lower resolution), copied into textures, and the shadows can be treated by exploiting texture bilinear interpolation. However, the basic approach may have incorrect results with specular highlights and may suffer with complex objects in terms of the performance capability.

2.4.3 Reducing the Complexity

Because of the large number of shadow polygons that are processed, there has been much research to significantly reduce such complexity. The optimizations can basically be categorized into three approaches that can be used in combination: a per-mesh silhouette approach, a global scene analysis optimization, and a hybrid shadow depth map approach.

These discussions are useful for both CPU and GPU versions of the shadow volume algorithm. On the GPU, this allows fill-rate reduction of the stencil buffer, which is crucial for performance.

Per-mesh Silhouette Approach

Many efficient implementations [46, 37, 48, 165] eliminate redundant shadow polygons within a single polygonal mesh by exploiting some silhouette identification

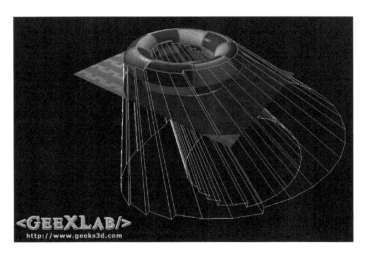

Figure 2.26. Reduced number of shadow polygons generated as a result of the object silhouette. *Image courtesy of Jerome Guinot/Geeks3D.com.*

(see Figure 2.26, where shadow polygons are generated only at the silhouette of the object). These silhouettes mainly correspond to the polygon edges that have different $\hat{N} \cdot \hat{L}$ positive/negative values (the view-dependent boundary case) and also to edges that have no other shared edges (the view-independent boundary case). Such polygon edges are the only edges that require the generation of shadow polygons, i.e., any internal polygon edges do not need shadow-polygon generation. Such an approach can potentially reduce the number of shadow polygons quite significantly.

As a side note, two observations about silhouettes are interesting to mention. First, the degree of a silhouette vertex is even [6], where the degree n means the vertex on such a silhouette can be connected by n silhouette edges. Second, another observation from McGuire [399] is that in many triangle meshes consisting of f triangles, the number of silhouette edges is approximately $f^{0.8}$.

Note that the above silhouette optimizations are only accurate for 2-manifold shadow casters, and there have been attempts at improving the generality of the meshes [46, 8, 400, 306]. Bergeron [46] indicates that silhouette edges with two adjacent triangles should increment/decrement the shadow count by 2 and increment/decrement the shadow count by 1 for open edges. To generalize this approach even more, Aldridge and Wood [8] and Kim et al. [306] compute the multiplicity of each shadow polygon, and the shadow count is incremented/decremented by this multiplicity value (although the usual case is either 2 or 1, as in the case identified by Bergeron). While supporting non-manifold shadow casters, these approaches require storage of the multiplicity value per shadow polygon, and a GPU stencil buffer implementation is more complicated because the standard stencil buffer only supports incrementing/decrementing by 1; additionally, the risk of exceeding the stencil buffer's 8-bit limit (shadow counts above 255) becomes higher. To mitigate the above issues, for the case of increments/decrements of 2, creating double-quads is a possible brute-force approach. Another possibility is proposed by McGuire [400], who experiments with additive blending to a color buffer.

The shadow count also needs to consider some portions of the original mesh (i.e., the light cap) within the shadow volume polygons. However, those polygons can be culled because of their irrelevance to the shadow-count computations [345, 397]. For example, if the occluder is between the light and the viewpoint, and the view direction is pointing away from the occluder, these polygons are not necessary for consideration for the shadow count. Similarly, if the viewpoint is between the occluder and the light and the view direction is pointing away from the occluder, then these polygons do not need to be considered for the shadow-count computation. In fact, in the latter case, all the shadow (silhouette) polygons for this mesh can be culled from any shadow-count computation. This work can reduce the fill rate for GPU-based solutions even more.

Global Scene Analysis Optimizations

Beyond optimizing for just a single mesh, conservative occlusion culling techniques as discussed by Cohen-Or et al. [106] can be used as a global scene analysis optimization technique. Additional specific examples are elaborated below.

Slater [537] points out that a shadow volume completely enclosed within another shadow volume can be eliminated. In other words, by processing shadow volumes in a front-to-back order from the light source, simple configurations of shadow volume clipping can reduce the extent of shadow volumes closer to the light source. This can be seen in Figure 2.27, where the shadow volume of occluder A can eliminate the need for the shadow volume for occluder B as a result from this sorting.

Fill rates can be further improved by reducing the size of the shadow polygons. For example, the intensity of the light diminishes as the square of its distance; thus, shadow polygons can be clipped at a distance where light intensity becomes so small that it does not affect shading [345, 397]. Similarly, the shadow polygons can be clipped beyond a certain camera depth range [398] because those points are not easily visible. Both techniques may result in shadow errors if there are many shadow-casting lights, and the accumulation errors may add up (see Section 2.5.3 for additional implications of handling many lights).

Another reduction of shadow polygon size can be achieved by clipping the shadow polygons to exactly fit to camera-visible objects [357, 155], a tool referred to as scissors [346]. A bounding volume hierarchy can also be traversed front-to-back to accelerate the determination [631, 565] of the pruning suggested in [537, 357, 155] because entire nodes (in the hierarchical bounding volumes) can be pruned along with all the leaf nodes that reside under the node.

In another optimization, a hierarchical shadow volume algorithm [3] is presented, where the screen is divided up into many tiles of 8×8 pixels. If the shadow polygons do not intersect the bounding box formed by the objects within this 8×8 tile, then the points within this 8×8 tile are either entirely in shadow or entirely in light, and the fill rate can be reduced because the stencil value for this 8×8 tile will be the same, and can be computed with any arbitrary ray through the 8×8 tile.

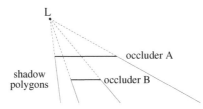

Figure 2.27. Occluder A can eliminate the need for the enclosed shadow volume from occluder B.

Fawad [169] proposes a lower level of detail in certain cases for the silhouette computation, thus reducing the time needed for the computation of the silhouette as well as the complexity of the number of shadow polygons. In addition, temporal coherence is applied to the next frame to further reduce the silhouette computation time. While this sounds theoretically interesting, self-shadowing is likely going to be a difficult problem to get around due to the different levels of detail. Zioma [675] provides a partial solution for closed surfaces, where front-facing (to the light) polygons are avoided for near capping, although self-shadowing concerns remain at the silhouette polygons.

Hybrid Shadow Depth Map

An interesting contribution in reducing the complexity of the general shadow volume approach comes from McCool [393]. A standard shadow depth map is first created, and then an adapted 2D edge detector is applied on the shadow depth map to identify the edges that form the only polygons casting shadow volumes. This scheme results in no overlapping shadow volumes (which others [37, 605] have tried to achieve through geometric clipping of volumes, but the computations are very numerically unstable); thus, only a 1-bit stencil buffer on the GPU is necessary to process the shadow volumes. When visualizing static scenes (i.e., the shadow depth map and edge detector preprocessing are not needed per frame), the actual shadow volume rendering should be very fast. However, the extra overhead in converting the shadow depth map to shadow volumes via an edge detector can be slow. The shadow depth map resolution limits how well these extracted edges capture smaller details and can lead to some aliasing artifacts in the resulting shadow volumes.

In the hybrid shadow-rendering algorithm proposed by Chan and Durand [84], the shadow depth map is initially generated. For the non-silhouette regions, it can be determined whether the region is entirely in shadow or entirely lit. For the silhouette regions, shadow volumes are used to determine exact shadowing, thus significantly reducing the fill rates because only silhouette regions are being considered. Unfortunately, this algorithm also has the same concerns as the above approach by McCool.

2.4.4   Shadow Volumes Integrating BSP Trees

Another variation of shadow volumes employs the combination of shadow volumes and binary space partitioning (BSP) trees [91]. A BSP tree is constructed that represents the shadow volume of the polygons facing the light. The shadow determination is computed by filtering down this shadow volume BSP (known as SVBSP) tree. This variation is quite different from the standard shadow volume approach because there is really no shadow incrementing nor decrementing, and it also has the advantage that there is no need to compute an initial shadow count. While speed improvements of the approach have been attempted [356], it has not

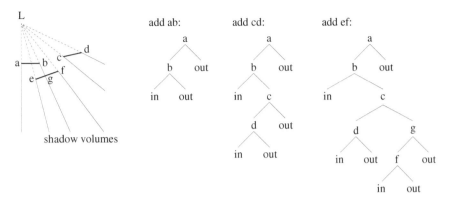

Figure 2.28. Construction of SVBSP so that tree traversal can be done to determine if a point P is in shadow.

been attempted on the GPU and likely remains slower than the regular shadow volume approach, although it may be efficient for computing shadows of indoor environments, with simpler and fewer openings for light to come through.

Figure 2.28 illustrates what the SVBSP tree looks like with the insertion of polygons (edges) ab, cd, and ef, respectively. Note that clipping is needed to ensure nonoverlapping cases, which means numerical stability issues must be handled with care. For a particular point P to be shaded, its shadow status can be determined by traversing this tree—left and right traversal is determined where P is with respect to the slices (at each node).

In terms of dynamic scenes, Chrysanthou and Slater [99] employ a BSP tree and a shadow-tiling approach (in the form of a hemicube) to get quick updates in dynamically changing scenes. Chrysanthou and Slater [100] then extend the SVBSP tree [91] to get dynamic updates in dynamic scenes to quickly add and delete changing elements of the scene, as well as to merge the modified SVBSP tree. Batagelo and Junior [37] use the SVBSP to prune the scene and only use the relevant shadow polygons within the standard stencil buffer.

2.5   Ray Tracing

The basic ray tracing approach to compute shadows is very simple. Most of the research has been focused on speeding up the ray tracing approach, although most methods remain generally inappropriate for real-time applications (except perhaps those covered in Section 2.5.2). The methods of acceleration include shadow culling algorithms (Section 2.5.1), combining ray packing and modern architectures (Section 2.5.2), dealing with many lights (Section 2.5.3), and speeding up antialiasing (Section 2.5.4).

2.5.1   Shadow Culling Algorithms

There are many intersection-culling algorithms to accelerate ray tracing, and many have been discussed in detail in other references [198, 526]. Of particular interest for shadow algorithms is the light buffer, hybrid shadow testing, voxel occlusion testing, and conservative shadow depth maps. Most of these algorithms are mainly good for static scenes or offline rendering and are highlighted because parts of their algorithms are also useful for progression towards real-time purposes.

Light Buffer

The light buffer [219] consists of six grid planes forming a box surrounding the point light source (see Figure 2.29). Each cell of the buffer contains information on the closest full occlusion from the point light (the closest object that occludes the entire cell) and sorted approximate depth values of candidate occlusion objects that project in the cell. For each point to be shaded, if the depth value of its corresponding light buffer cell is greater than the closest full occlusion distance, then the point is in shadow. Otherwise, shadow determination requires intersection tests

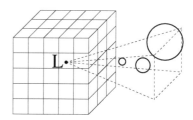

Figure 2.29. The light buffer of a point light source. Each pixel of the light buffer contains a depth-sorted list of encountered objects up to an object producing full occlusion.

with candidate occlusion objects of the cell. They are performed in order with respect to the depth values (starting with the object closest to the point light) until either an intersection hit is found (in shadow) or the depth value of the candidate occluding object is greater than the intersection point (not in shadow). This data structure has been employed in some variation of shadow volume algorithms [537] (Section 2.4.1), radiosity computations [104], and ray tracing soft shadows [407]. Also note that the image in Figure 1.16 is rendered using this light-buffer approach.

A ZZ-buffer [494] is a different data structure, but can be used in a very similar fashion to the light buffer to accelerate object intersection with shadow rays. The ZZ-buffer is basically a Z-buffer with each pixel containing pointers to geometry information (geometry that resides in that pixel). Thus, a rasterization step to populate the ZZ-buffer is done per light; then for each shadow ray, only the objects in the ZZ-buffer's projected pixel are intersected.

Hybrid Shadow Testing

Hybrid shadow testing [164] uses a voxel data structure to store the shadow polygons as invisible surfaces (as in shadow volumes of Section 2.4). The shadow count is updated as the ray traverses the voxels. No shadow rays need to be generated for this scheme, but intersection calculations with the shadow polygons in traversed voxels are necessary. When the closest intersected surface is found, the shadow count is checked. If the count is 0, then the surface is not in shadow; otherwise the surface is in shadow. However, since it might need to deal with a large number of shadow polygons, the implementation resorts to the traditional shadow ray approach under such circumstances.

Voxel Occlusion Testing

For voxel occlusion testing [638], each voxel uses up an extra 2 bits per light source. Its value indicates the level of opaque occlusion of the voxel with respect to each light source: full occlusion, null occlusion, and unknown occlusion (see Figure 2.30). The shadow umbra of all objects is scan-converted into the voxel occlusion bits, such that if the voxel entirely resides inside or outside the shadow umbra, then the voxel occlusion value is full or null occlusion, respectively. If the voxel contains the boundary of the shadow umbra, then it is marked with unknown occlusion. When the viewing ray intersects an object in a voxel that has either full or null occlusion, then any intersected object in the voxel must be in shadow, and no shadow rays need to be generated. If unknown occlusion is found instead, then the fastest method of shadow determination is to resort back to voxel traversal of the shadow ray. However, as each voxel is traversed, its occlusion value is checked—a quick exit condition for the shadow ray traversal is available if any of those traversed voxels show a full or null occlusion value, which will again mean that the point to be shaded is fully shadowed or fully lit, respectively.

The objective of this approach is to get known occlusion status as often as possible, and thus for polygonal meshes, it is best to preprocess voxel occlusion testing with the per-mesh silhouette optimizations described in Section 2.4.3 instead of processing on a per-polygon basis.

For interactive applications, a high enough voxel resolution may be used to reduce the occurrence of unknown occluded voxels, or it can be used to encode only totally occluded or unoccluded voxels [650, 503] (Sections 3.7.4 and 2.6,

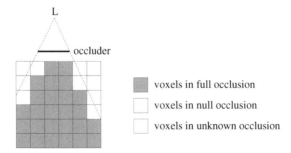

Figure 2.30. In voxel occlusion testing, each voxel of the regular grid is marked as full, null, or unknown occlusion from one light source.

respectively). Such an encoding can also lead to the ability to generate atmospheric shadows as well (Section 5.6.1).

An alternative but similar optimization is introduced by Djeu et al. [133]. Objects are assumed to be closed and totally opaque. An interior fill operation of the objects is done to fill up the k-d tree. When computing shadows, the shadow ray traverses the k-d tree, but instead of first intersecting the mesh, the shadow ray first tries to locate filled regions. If a filled region is found, then it is in shadow. If a filled region is not found, then costly intersection calculations with the mesh are needed, but this is usually only required at the silhouette of the object (from the perspective of the light).

Conservative Shadow Depth Maps

A couple of independent papers [337, 243] make the observation that shadow depth maps usually provide a decent approximation to the shadowing occlusion and can be used as a first pass to avoid shooting lots of shadow rays (in the standard case) and to make the final quality higher. The shadow depth maps created in the first pass tend to be more conservative, to cover cases where the shadow occlusion is uncertain. Whether or not to shoot a shadow ray is determined on a view-pixel basis—these shadow rays are often needed at the silhouette of the geometry.

The above two papers differ in their approaches for determining which pixels need to shoot a shadow ray. Hertel et al. [243] store in the shadow depth map the minimum z-value of the entire coverage of triangles in the shadow depth map pixel. A check is also done if the shadow depth map pixel is entirely covered. If entirely covered, then the shadow-ray origin compared to the shadow depth map pixel z-value can determine whether the point is in shadow or lit without shooting a shadow ray. Otherwise, a shadow ray is shot, but traversal of the shadow ray does not need to go beyond the shadow depth map pixel z-value (i.e., early ray termination).

Lauterbach et al. [337] tag each shadow depth map pixel as requiring a shadow ray to be shot if the Z-depth pixel differs significantly from the neighboring Z-depth pixels. In essence, it is tagging the silhouette to discover the need for a shadow ray.

Both papers test their approaches on the GPU. It is unclear how effective the algorithms would be when insufficient shadow depth map resolution occurs.

2.5.2   Combining Ray Packing and Modern Architectures

Relating to the previous discussions on shadow-culling algorithms, other papers [453, 641, 94, 545] accelerate the computation of shadows from simple lights using voxels as the culling approach. They have not enjoyed as much success as hoped because the performance bottleneck in an optimized ray tracer no longer resides with the reduction of floating-point computations and object intersection tests. Rather, lack of locality of memory references has become the bottleneck with modern platforms. This is precisely why a ray-packing approach towards ray tracing has met with much more success [610, 613, 64] in improving performance. Ray packing generally means doing computations for a group of rays such that memory accesses only the same, small set of geometries.

Ray packets, in conjunction with a suitable culling data structure (bounding volume hierarchies (BVH), k-d trees, BSP trees, uniform grids, concepts all described in [198, 526]), the SIMD architecture (e.g., Intel's SSE), and a highly parallel environment in either CPU [610, 613] or GPU [248, 213, 464, 676], have enjoyed significant speed improvements. However, the memory requirements remain large, and texturing is currently not well supported in the above environments.

To help with the memory requirements, out-of-core algorithms [461, 609, 611, 73] have been introduced. The order of ray computations is done differently to optimize the above ray packing in conjunction with voxel cache hits. This allows for much less data set I/O to move in and out of the entire data set.

2.5.3   Dealing with Many Lights

In scenes that contain many shadow-casting lights, it becomes quite costly to compute shadows for each point to each of those lights. To achieve a similar-looking result, a number of algorithms have been proposed.

Bergeron [46] defines a sphere of influence around each point light source, for which the radius is related to the light intensity. Any point outside the sphere is not considered for shadowing. While efficient, the method fails when many light sources of low intensity are ignored, but their combined contribution could be observable. This is also the case when approximating an extended light source with many point lights. Using more light samples would reduce the overall scene illumination and therefore also the shadowing effects.

Ward [625] determines, among all the lights, an ordering of which lights contribute most to the current point to be shaded. He only calculates lighting and shadowing based on the lights that have significant contributions. As a result, shadow rays only need to be shot to some small number of lights per point to be shaded.

Shirley et al. [527, 528] attempt to find good probability density functions using Monte Carlo integration for direct lighting of extended light sources (for soft shadows) as well as for a large number of lights. The algorithm subdivides the scene into voxels, and for each voxel, separates the set of lights into either an important or an unimportant set. While each light in the important set is sampled by shooting shadow rays per light, only one arbitrary light from the unimportant set is chosen for the shadow ray shooting to represent the entire unimportant set of lights. The use of exactly one shadow ray provides good results because the noise generated from low-intensity changes tend to be less objectionable over an animated sequence. The optimal number of shadow rays to achieve accurate enough quality has not been researched in this work. Figure 2.31 shows one example from this approach.

Fernandez et al. [171] also subdivide the scene into voxels, more specifically octrees. The basic preprocessing results in a voxel indicating whether the entire space occupied by the voxel is fully occluded, partially occluded, or unoccluded for each light. The storage is quite large: for each light, fully occluded voxels store nothing, unoccluded voxels store a pointer to the light, and partially occluded voxels store a pointer to the light and links to potential occluders for that voxel. During the shading phase, for fully occluded voxels, nothing needs to be done; for unoccluded voxels, the lighting equation is computed; for partially occluded voxels, shadow rays are shot to determine the shadowing. To accelerate the case of many lights, Weber's law, $K = \Delta I / I$, is applied to unoccluded and partially occluded voxels. Weber's fraction K determines which points to be shaded do not really matter from visual perceptions. Weber's law therefore identifies lighting that could be eliminated from computation based on quick evaluations of whether each lighting case per voxel contributes noticeably to the final shading.

Figure 2.31. Results with reduced sampling of many lights. *Image courtesy of Changyaw Wang.*

The many-lights problem might not come only from multiple emitting light sources. Virtual point lights (VPL) can be generated to propagate indirect illumination in global-illumination algorithms. Techniques derived from instant radiosity [299], lightcuts [614], and light transport matrices [229, 443] are discussed in more detail in a section about VPL (Section 5.10.2) in the global illumination section (Section 5.10).

Note that although the above work has been done with respect to ray tracing, it can be extended to some other shadow algorithms as well, as the point of the above work is to avoid shadow computations for each point for all lights. In addition, although the above research work is very interesting, it has not been used in production work as far as we know. In production work, in order to save rendering time, there may be a lot of lights, but only a few lights are tagged for shadow calculations. Also, options such as associating certain objects with certain shadow-casting lights (so that the object casts shadows only by those lights, to reduce rendering computations) are available in most software renderers. Finally, there are also options to select certain objects as shadow casters (or not) or as shadow receivers (or not)—i.e., not all objects need to be considered for shadowing.

2.5.4   Antialiased Shadows

To get better antialiased (hard) shadows without shooting a large number of shadow rays all the time, some cone tracing [10] (Section 4.7.2) or beam tracing [231] extensions have been implemented. For example, Genetti and Gordon [188] use cone tracing [10] not to determine visibility, but as an indication of how many shadow rays need to be shot and in which section of the cone to get good antialiasing. The criteria of the presence of different polygons is used to determine how complex that cone region might be and thus indicates how many rays need to be shot. This may be ideal for parametric surfaces, implicit surfaces, or large polygons, but not ideal for polygonal meshes, as polygonal meshes will always require many shadow rays to be shot based on the above criteria. Similarly, Ghazanfarpour and Hasenfratz [193] use beam tracing [231] to get antialiased shadows. It encapsulates the different cases in which the objects will intersect the beam. This approach is again mainly ideal for large polygons but not polygonal meshes because the meshes will slow down the approach.

Another fast antialiased shadow algorithm involves acceleration based on information from the current pixel's sample results [643]. That implementation assumes antialiasing is achieved by adaptive supersampling [629]. Instead of always shooting a shadow ray when extra samples are needed (due to some aliasing), the ray trees of the surrounding samples are first checked. Additional shadow rays are shot only if the invocation of the extra samples is due to different shadow occlusion values. Otherwise, the extra sample's shadow occlusion value is assumed to be the same as the surrounding samples. Note that this can provide important performance gains if significant aliasing is caused by geometry silhouettes, noisy textures, specular highlights, etc., and not shadow aliasing. For example, looking at Figure 1.16, one would assume that a single shadow ray per pixel is shot when rendering the counter top, except at the border of the shadow. Due to the textures on the counter top, additional subpixel rays are shot to antialias the texture, but the shadow information should not need to be computed.

2.6   Other Hard Shadow Algorithms

A number of algorithms have been investigated for the specific cases of convex polygonal meshes [618], shadow outline processing as used in the video game Outcast [500], discretized shadow volumes in angular coordinates [585], theoretical studies of precision shadows through epsilon visibility [146], and subtractive shadows to generate shadow levels of detail [124].

An interesting class of algorithms use GPU projection textures [426, 421, 438] to render shadows. Nguyen [426] renders projected polygons to a GPU texture and applies this texture onto the receivers by computing its projective texture coordinates on the fly. Two papers [421, 438] discuss the use of shadow depth map results to project shadow textures onto receivers. The shadows are generated as a result of multipass rendering of the shadow textures with the receivers. Although these techniques are quite fast, the main limitations are that an object cannot self-shadow and can only be an occluder or a receiver, not both. Further, it is not feasible to implement this for more than one shadow-casting light. Oh et al. [437] combine projection textures with shadow depth maps to alleviate the above limitations.

Schaufler et al. [503] implement a variation of voxel occlusion testing [639] using an octree. Because the lights and models are not changing during the visualization, the fixed shadow occlusion values can be embedded in voxels to indicate whether a certain voxel region is in shadow or not. They also improve the detection of full occlusion due to combined occlusions (occluder fusion) of disjoint occluders. When the voxel contains regions that are in shadow and some region not in shadow, it requires additional shadow rays to be shot to determine the shadow occlusion.

There is also an entire class of algorithms focusing on out-of-core approaches to reduce memory of large data sets. A small discussion was provided in Section 2.5.2 specific to ray tracing. A more generalized discussion of out-of-core algorithms can be found in the following reference [531].

2.7   Trends and Analysis

In reviewing the contents of this chapter, it is clear that there is no algorithm for all situations. From a research perspective, shadow (depth) map papers have been the most abundant in an attempt to automate the correction of shadow artifacts. The current count for the papers are

Planar receivers of shadows	5
Shadow (depth) map	103
Shadow volumes	44
Ray tracing	37

2.7.1   Offline Rendering Trends

In offline rendering situations, the trend is that most turnkey software (such as film and video software including Houdini, Renderman, Maya, Mental Ray, Lightwave, etc., and CAD software including SolidWorks, AutoCAD, TurboCAD, etc.) provide shadow options from both ray tracing and shadow depth maps, and shadow volumes are rarely offered. Shadow depth maps offer the faster, less memory consuming (unless out-of-core techniques are applied (see Section 2.5.2)), approximate, and usually aesthetically pleasing quality solution, whereas ray tracing provides the most flexible and exact solutions, usually used when shadow depth maps become troublesome or fail. Also, with platforms supporting high parallelism (on the CPU or GPU) or scenes with many shadow-casting lights (due to less per-light preprocessing or many light optimizations (see Section 2.5.3)), ray tracing is becoming more affordable.

Shadow depth maps are much more appropriate than shadow volumes in offline rendering situations because shadow depth maps allow efficient handling of large data sets, higher quality renderings due to user-input tweaking (and the users' willingness to tweak parameters), and percentage-closer filtering capabilities to achieve fake soft shadows, which can also mask the straight edges of polygons that shadow volumes cannot. As well, shadow depth maps can deal much more effectively with the terminator problem (Section 1.4.1). Finally, shadow depth maps can easily integrate shadows from any surface type with the flexibility of using different visibility determination algorithms for each surface type (e.g., Z-buffer for polygons, ray tracing for algebraic surfaces, simplified splatting for voxels, etc.).

The most common implementation of shadow depth maps in offline rendering situations is the implementation described by Reeves et al. [476]. Some offer the mid-distance solution [642] for avoiding bad self-shadowing, and many offer some tile-like implementations [172, 333, 480] to reduce memory usage. It is rare to see any nonlinear shadow depth maps implemented, however. This is likely due to the lack of image-artifact-free robustness (especially with aliasing far-away geometry) of the nonlinear approaches in many offline rendering situations as well as the ability of the production users to optimize their scenes very well (i.e., the entire model is rarely submitted for rendering of specific scenes; parts of the models are manually culled) that allows linear shadow depth maps to be very effective.

2.7.2   Real-Time Rendering Trends

In real-time situations, it is rare to see ray tracing in turnkey real-time engines. Although ray tracing is, by far, the most flexible, unfortunately it remains inappropriate if real-time frame rates are desired unless the platforms support high parallelism (on either the CPU or GPU), which is not universally true yet. Shadow depth map and shadow volume algorithms tend to be much faster on common hardware. Shadow volumes are quite at home in real-time, GPU-based, simple environments (e.g., Sketchup) because the polygon count (for shadow-casting objects) tends to be small, polygons are the dominant primitive for real-time applications, and shadow volumes can deal with polygons very efficiently and accurately with little to no user input. However, shadow volumes have not been used in many AAA titled games (high-quality games with large budgets) or other real-time environments because a consistent frame rate cannot be achieved despite the

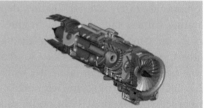

Figure 2.32. A closed model (left) is no longer closed after cross sectioning (right). *Image courtesy of NGRAIN (Canada) Corporation.*

optimizations discussed in Section 2.4.3. Also, note that some important shadow volume optimizations (Section 2.4.3) require that objects be closed, but this assumption may not be true upon creation, or the assumption may be broken due to real-time interactions such as cross sectioning (without capping). See Figure 2.32 for a simple example of a closed model (left) not being closed after cross sectioning (right). However, new approaches such as the per-triangle shadow volumes [536] do show some interesting promise.

On the other hand, shadow depth maps have been quite abundant in real-time engines for more complex environments (such as Second Life, Unity 3D, DX Studio, Renderware, etc.). In fact, most implementations involve either warping shadow depth maps or z-partitioning variations (Section 2.3.7). This has become important to allow entire worlds to be rendered with decent shadow quality without resorting to very high shadow depth map resolutions. Z-partitioning variations seem to be gaining more momentum because they generate better quality, tend to be more artifact free, and the quality/performance effects can be controlled by allowing the user choice of the number of split planes. In fact, z-partitioning variations seem to be the approach of choice for AAA titled games. There has also been a trend to combine z-partitioning with variance shadow maps to get very good-quality renderings. This is not surprising considering that the two approaches are easy to code and combine.

If only shadows cast on a floor/wall in real time are required, the planar receivers of shadows approach [57, 567, 223] is clearly the winner and most often used implementation under such circumstances (e.g., WebGL). It is even appropriate and easy to combine approaches: shadows on a floor (where some floors can be a significant size/portion of the scene) are computed using the planar receivers of shadow approach, and the rest of the scene is computed using shadow depth maps, for example.

2.7.3   General Comments

Figure 2.33 shows a decision tree for novice practitioners that considers which approaches will work in certain scenarios. Note that the image quality aspect is ad-

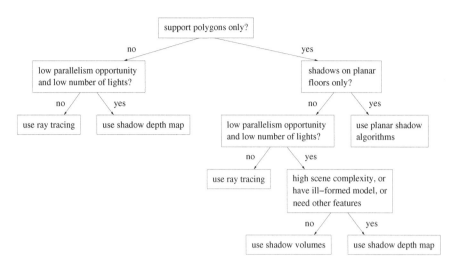

Figure 2.33. Graph for determining which shadow approach to use.

dressed at the bottom of the tree, which may be surprising but not unexpected, because the external considerations of the entire system (of which shadow algorithms are a part) are generally more constraining.

Pluginable shaders and lights have become commonplace in high-quality renderers, such as Renderman, Mental Ray, Maya, etc. This is because shaders and lights only require localized information such as intersection point, texture coordinates, tangent values, etc. Allowing shadows to be easily pluginable is tougher however. In the case of ray tracing, this is simple because the user can write user code to add or change shadow rays during shading. However, in non ray tracing algorithms, pluginable shadow algorithms are not that simple because shadow algorithms are basically some variation of the visibility determination algorithms. As such, shadow algorithms need access to the data structures of the geometry. Thus, the information required is not so localized. In the case of Renderman, the main support for shadow algorithms is the shadow depth map, and it is treated as a separate rendering pass (Z-depth render mode); then, some user code incorporates the comparison of depth values during shading.

And that is just the literature on hard shadows. The discussions on soft shadows are even more complex and its algorithms less organized. In fact, if the reader only needs to consider polygons, then the recommendation is to skip the next chapter for now and proceed to Chapter 4 on soft shadows. However, if non-polygons need to be considered, the following chapter should be read next, as it covers hard shadows for non-polygonal objects.

CHAPTER 3
Supporting Shadows for Other Geometry Types

3.1   What's Covered in This Chapter

This chapter covers algorithms that generate shadows for non-polygonal primitives. Most of the published papers for such primitives discuss hard shadows only with very few delving into soft-shadow discussions. The list of non-polygonal primitives discussed in this chapter includes

- Higher-order surfaces, such as parametric, implicit, subdivision, and CSG (constructive solid geometry) surfaces (Section 3.2).

- Image-based rendering supporting impostors (Section 3.3) [274].

- Geometry images (Section 3.4).

- Particle systems (Section 3.5).

- Point clouds (Section 3.6).

- Voxels within volume graphics (Section 3.7) [349, 160, 385].

- Heightfields (Section 3.8).

Note that higher-order surfaces, particle systems, and point clouds are usually natively created in some software (e.g., parametric surfaces created in Maya), whereas heightfields, voxels, and impostors tend to be natively created in other forms and then converted to their particular geometry type for convenience of rendering and visualization.

3.2   Higher-Order Surfaces

Higher-order surfaces, such as parametric surfaces, implicit surfaces, and CSG, are critical to computer graphics because they represent exact mathematical representations of the 3D model and can be described with less data input (thus using less memory) than the many vertices for polygonal meshes. There are multiple ways to render higher-order surfaces: the most popular among them use numerical iteration (Section 3.2.1) or rely on a form of polygonization (Section 3.2.2). There are also a few variants that use shadow volumes to directly generate shadows (Section 3.2.3).

Before proceeding, we provide a few words on what these higher-order surfaces are:

- *Parametric surface*, also known as a spline surface, which is a particular and popular form of a polynomial parametric surface. A parametric surface in 3D is defined by a function of two parameters, usually denoted by (u, v). The surface is defined by a set of user-input control 3D points and the interpolation function for those points (see Figure 3.1). The most common example of parametric surface is the cubic NURBS.

- *Implicit surface*, also known as a level-set or isosurface. An implicit surface is defined by a mathematical function F, where $F(x, y, z) \leq c$. For example, a unit sphere located at (0,0,0) is defined by $F(x, y, z) = x^2 + y^2 + z^2$. A variation of an implicit function is known as algebraic surface, metaballs, or blobbies, with the definition that $\sum_i F_i(x, y, z) \leq c$. An example of blobbies can be seen in Figure 3.2, where the water is composed of blobbies and they are rendered using numerical iteration techniques.

- *Subdivision surface*, which starts from an initial polygonal mesh that is recursively refined by inserting more vertices and faces to replace the ones from

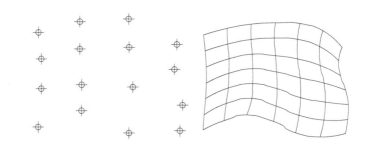

Figure 3.1. Control points on the left, with the piecewise polynomial definition, result in the surface generated on the right.

Figure 3.2. Blobbies and their corresponding shadows, by Duncan Brinsmead. *Image courtesy of Autodesk, Inc. © 2007. All rights reserved.*

the previous iteration. The refinement schemes compute new vertex positions within a given neighborhood; they are interpolating when the original vertices are matched with the subdivision, and otherwise approximating.

- *Constructive solid geometry (CSG)*, procedural modeling technique that uses Boolean operators (such as union, intersection, difference) to combine objects.

Oddly enough, we found little to no literature on shadow algorithms for boundary representation (B-rep) surfaces. We suspect this is because the rendering of those surfaces is almost always done through polygonization. Further, B-reps are usually applied in CAD situations where shadows are not as critical. Thus, we have no coverage of B-rep surfaces as higher-order surfaces in the upcoming discussion.

3.2.1   Numerical Iteration

Rendering of higher-order surfaces can be achieved through direct rendering that applies numerical iteration techniques. Such techniques compute the intersection of a 3D ray with the surface, and the closest intersection point with the ray is determined. Numerical iteration techniques require a starting guess where the closest point might be; then, iterations of the guesses hopefully result in the convergence of the solution. The quest for all numerical iteration techniques is to bound the intersection point's range sufficiently to allow both fast and correct convergence.

Example numerical iteration techniques for implicit surfaces can be found using Newton's iteration and regula falsi [55], deriving analytic solutions to low-order polynomials [224], and using interval arithmetic [285, 411]. To achieve faster performance, the following papers use GPU to directly render implicit surfaces [363, 588, 287, 532].

Example numerical iteration techniques for parametric surfaces can be found in the following: using Laguerre's root-finding algorithms [281, 432], multivariate Newton's method [572], interval arithmetic [591], subdividing polygonal approximation [644], Bézier clipping [432], and Chebyshev polynomials [180]. To achieve faster performance, the following papers use GPU to directly render parametric surfaces [216, 448, 286, 318].

Although the above numerical iteration techniques hint at a ray tracing approach for shadowing, and while it is definitely the case, the intersection tests can be done as a preprocess to fill up depth information for shadow depth maps as well. Initial guesses can then rely on neighboring shadow depth map pixels.

The numerical robustness to guarantee closest intersection hits with the earlier approaches can be difficult to achieve, resulting in potential holes in the rendered images. However, robustness has been increasing. Further, rendering speed may be slow due to poor convergence rate to evaluate a single point visibility but significantly improved due to GPU-based algorithms. The quest remains high for direct rendering of higher-order surfaces due to the small amount of memory needed and exact intersections achieved, as compared to polygonization of the surfaces, which will be discussed next.

3.2.2   Polygonization

Instead of numerical techniques to deal with the same parametric surfaces and implicit surfaces, polygonization (tessellation of the higher-order surface into a polygonal mesh) is another approach. However, polygonization has its costs, including the need for larger amounts of memory to deal with the polygonal mesh, the need to approximate a smooth surface with a sufficient number of polygons, and the need to deal with the terminator problem (as described in Section 1.4.1, although this problem can be reduced with finer tessellation).

There is a variant of polygonization that avoids some of the above concerns. In the Renderman implementation [110], fine tessellation (micropolygons) occurs on a per-tile basis, where each tile is a small region of the entire Z-buffer. The tessellation is thrown away once that tile has completed rendering. In such a case, neither memory nor terminator problems are of concern. However, this only works for a Z-buffer rendering approach, and care must be taken to account for the differing levels of detail of the tessellation between the Z-buffer render and the shadow depth map render, resulting in improper self-shadows (this is discussed in Section 2.3.2).

Example references on polygonization from implicit surfaces [59, 58] and parametric surfaces [607, 523] are available. Many other geometry types also employ polygonization for the final rendering and shadowing steps due to some subset of the above reasons, such as Marching Cubes [367] to convert voxel data to polygons, and techniques to convert point-cloud data to polygons [594, 552]. Direct shadow rendering algorithms for those other geometry types are discussed in sections starting with Section 3.6.

Note again that if the implementation employs polygonization, then the algorithms in Chapter 2 are more than relevant to the reader, and the graph of Figure 2.33 to determine which algorithms to use applies.

3.2.3   Shadow Volumes

Jansen and van der Zalm [266] use shadow volumes to calculate shadows for CSG objects. Shadow volumes are first calculated for the shadow-generating parts of the silhouette (from the viewpoint of the light source) of each CSG object. A very large shadow CSG tree per object is formed. They consider only the front-facing parts of the CSG object to simplify the complexity of the shadow CSG tree. The shadow trees tend to be quite large and are needed on a per-object and per-light basis.

Heflin and Elber [235] extend the shadow volume approach to support freeform surfaces. Once the silhouette from the viewpoint of the light source is identified, a set of trimmed surfaces are created to represent the shadow polygons of the shadow volume. The shadow count computations intersecting from these trimmed surfaces can be slow.

Tang et al. [580, 579] use the GPU to accelerate silhouette extraction in geometry images (Section 3.4) from subdivision surface-based models.

3.2.4   Trends and Analysis

Polygonization remains the more popular technique in industry (offline or realtime). Since it can apply any of the rendering and shadowing algorithms described in this book, the rendering can be fast because the polygonal mesh can be accelerated on the GPU, and most importantly, polygonization allows code maintenance to be much simpler since the rendering algorithm essentially only deals with one type of primitive (polygon) and the tessellation components for handling the different higher-order surfaces can be self-contained. The latter point is especially key for some software which offers multiple surface types, e.g., Maya supports modeling of NURBS, subdivision surfaces, and polygons.

However, with direct rendering solutions on the GPU, and the advantages of less memory usage plus avoidance of the terminator problem, direct rendering solutions should become more viable. The decreased memory factor makes native support ideal for current tablets, but the GPU capabilities of these devices may not be powerful enough. Further, the exact rendering of surface detail (through direct rendering solutions) is also ideal for certain domains that require such accuracy, such as CAD or scientific applications/simulations. Also, direct rendering is most efficient when using blobbies to represent particles (such as water), due to the large data set complexity.

Another scenario that is becoming more common is the need to deal with *distributed learning* scenarios (distributed over the internet), where the 3D files need to be quickly downloaded (via the internet) in order to visualize the content. With certain higher-order surfaces (e.g., NURBS), the geometry representation can be very compact and thus quite suitable for fast network transmission. As a result, being able to render the geometry natively, without the need to wait for the polygonization would be a significant advantage, i.e., perceived wait of download plus polygonization may be long and hurt the user experience.

3.3   Image-Based Rendering for Impostors

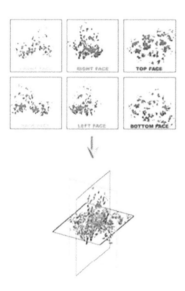

Figure 3.3. The combined effect of a few planar impostors can approximate the rendering of a complex 3D geometry, here, foliage, which appears different on both sides of each impostor. *Image courtesy of Lluch et al. [361].*

Image-based rendering techniques [88] make use of textured semitransparent polygons as *impostors* in a scene. These impostors can be rendered faster in general than the many detailed 3D geometries that they are replacing, mainly because each impostor is usually built from very few underlying polygons. The image-based rendering techniques allow such impostors to be used in certain far-away views, benefiting from texture filtering within the impostor surface, instead of facing the aliasing of rendering multiple small, distant polygons. Impostors can also be interpolated between different textures, thus simulating changing appearance under different view angles, under different light directions, or for animated objects (e.g., a walking sequence for a human encoded as a movie sequence). Impostors are particularly useful in urban environments because of the many buildings, trees, or humans (crowds) that can be faked almost seamlessly. They can also be used multiple times as instances in a same scene or combined or textured differently to introduce some variability. Finally, because they are built on simple polygons, impostors integrate well within any polygonal-based rendering system.

Much of the research in image-based rendering has been focused on the computation of the seamless view interpolation of the impostors, as well as the computational cost and access speed of many GPU textures representing the impostors. Impostors have been represented by 2D cross images called planar impostors [374], billboards [123], multilayer depth images [520], or volumetric billboards [121]. A single object can be represented by one or multiple impostors, and vice versa, an impostor can represent several different objects. Figure 3.3 illustrates the combined effect of a few planar impostors.

In the brief coverage that follows, we present the literature on the shadowing techniques used in image-based rendering, specifically for large numbers of buildings, trees, and humans (crowds), where the application of image-based rendering appears to be the most obvious. Note that the majority of the shadow algorithms are assumed to be shadows landing on a ground floor.

3.3.1 Trees

In a very traditional approach, Zhang et al. [666] use a set of three orthogonal billboards, each billboard complete with color and alpha channel information. Shadow projection onto the ground plane is achieved with the alpha channel information and a stencil buffer. The method is only valid for distant trees, and these trees must be very symmetrical due to the minimal set of billboards.

Max and Ohsaki [389] simply precompute Z-buffer images from different view directions around a sphere, and project the 3D points extracted from the closest Z-buffer directions to produce an image from an arbitrary viewpoint. The 3D points should provide a better depth complexity, compared to blending between flat billboards. Shadow determination for the rendered 3D points comes from a similar construction of a shadow depth map.

Zhang and Nakajima [660] experiment with shadowing of impostors from trees. Features of the trees are captured as irregular lines, and projection of the irregular lines onto the ground plane forms the basis of the shadow generation. Several methods [132, 291, 661, 659] apply computer vision techniques to extract shadow silhouettes from image-based objects to achieve shadow projections of the silhouette on a planar ground.

Meyer et al. [406] render many trees on a landscape by also using image-based rendering techniques. A hierarchy of bidirectional textures is used to store different LODs to represent the tree, and the appropriate LOD texture is sent to the GPU. Shadowing is accounted for in the following combinations: a hierarchy of visibility cube-maps is built where each tree's cube-map accounts for visibility of other trees and other geometries (e.g., landscape); self-shadowing of the tree onto itself is taken care of in the bidirectional texture functions (BTFs) and horizon-mapping techniques [391]; a shadow depth map [630] is used to incorporate all of the above.

Qin et al. [470] use several vertical and a few horizontal 2D buffers to store a sampled representation of the shading and shadowing information. Voxels are also precomputed to approximate the visibility of the hemisphere surrounding a single tree in order to capture skylight shadowing. They handle shadows of a sunlight cast by a tree on itself, on the ground, and on neighboring trees, as well as softer shadows due to skylights.

3.3.2   Crowds

Loscos et al. [370, 583] represent humans walking in virtual cities, in which each human is treated as an animated impostor. Shadows of these humans are cast onto the floor by projecting a colored shadow version of the animated impostor itself, using a method similar to the fake shadows of Blinn [57] (see Section 2.2). Shadowing of buildings onto the humans is achieved using a shadow depth map of the city, and computing its height coverage against the human im-

Figure 3.4. Crowds are rendered using impostors to reduce the complexity of many 3D models. *Image courtesy of Tecchia et al. [583]. Computer Graphics Forum ©2002 The Eurographics Association and Blackwell Publishing Ltd. Published by Blackwell Publishing, 9600 Garsington Road, Oxford OX4 2DQ, UK and 350 Main Street, Malden, MA 02148, USA. Reproduced by permission of the Eurographics Association and Blackwell Publishing Ltd.*

postor. Shadows are displayed using a shadow impostor (the impostor is composed of two parts: shadow and semitransparency) mapped onto the human impostor. An example of such crowd rendering is given in Figure 3.4. Dobbyn et al. [136] use instead a stencil-buffer approach for the shadows on the ground in order to remove z-fighting artifacts and to better blend in shadows on a textured ground.

In crowds lit by many static light sources (e.g., street lamp posts), shadowing can become a burden. Ryder and Day [493] combine each of the many shadow maps into a single large shadow map, thus reducing light context switches. The resolution of each shadow map is dependent on its distance in the image. Each light source has also a display list of all static scene elements within its radius of influence, and at run-time, a 3D grid is used to identify the impostors potentially lit by each light source. The impostors are augmented with depth per texel and rendered correctly with these depths in the shadow maps and in the image in order to cast accurate shadows and self-shadows.

3.3.3 Trends and Analysis

The real-time applications of image-based techniques in industry have been popular in architecture projects, virtual reconstruction, and population of heritage sites, synthetic avatars in interactive dialogs, molecular visualization, some forms of hybrid rendering for a large terrain (further away landscape rendered as an impostor), and obviously, within the video game industry. The latter has seen its use of impostors becoming mainstream for vegetation, animated particle-based effects, or captured video sequences to represent phenomena such as smoke, torches, rain, and large crowds.

Standard projection, shadow depth map, and ray tracing algorithms apply in many situations, but completely image-based representations or those from polygons augmented with image-based data require adaptations. For trees and crowds, the adaptations discussed previously form a good base, but no single solution proves perfect, and more work is still expected. A number of applications for offline rendering have appeared, notably for rendering tree impostors. In the case of multiple cross-plane impostors, these applications often give the ability to the user to select how impostors cast shadows.

Finally, one could draw similarities between impostors and other representations such as point clouds (Section 3.6), voxels (Section 3.7), and heightfields (Section 3.8). One could suspect that some solutions for shadowing in one representation could successfully be adapted to another.

3.4   Geometry Images

An object formed as a polygonal mesh often has a one-to-one parameterization associated with its surface, which can be used for texture mapping purposes. By unwrapping the entire polygonal mesh into a rectangle, thanks to its surface parameterization, the resulting encoding of the mesh can be sampled in this rectangle in order to reconstruct a 3D point interpolated from the mesh polygon it intersects. Regular sampling of this encoding, like an image with regularly spaced pixels, reconstructs an entire mesh from the original mesh. This corresponds to replacing the usual RGB coordinates of a pixel by XYZ coordinates of a 3D point, or any other sampled information such as the coordinates of the normal vector at this location. This regular structure, called *geometry image* by Gu et al. [206], encodes vertex adjacency implicitly and offers several advantages for image processing operations that are more GPU-friendly. One such operation subsamples the geometry image to generate LOD representations for the mesh. Figure 3.5 shows a 3D mesh resampled from its geometry image representation. Several other uses and improvements of geometry images have been presented over the years.

Most shadow depth map or shadow volume algorithms apply trivially to a mesh resulting from geometry images. However, because image processing can be applied directly on a mesh extracted from geometry images, shadow volume algorithms appear more adapted to this representation. Therefore a few improvements have been exploited for shadow volume algorithms on the GPU.

Tang et al. [580, 579] detect the silhouette edges from the geometry image of a subdivision surface, store these edges in a silhouette image, and extrude on-the-fly quadrilaterals to construct the shadow volumes.

Fawad [170] precomputes LODs of the geometry images and the associated shadow volumes that are also stored as geometry images. The rendering gains are directly related to the reduced number of triangles rendered because the LODs for both geometry and shadow volumes are dynamically selected.

Figure 3.5. From left to right: original mesh, 3D coordinates of the mesh vertices encoded in RGB, 3D coordinates of the normals, and lower resolution resampled mesh from the geometry image. *Image courtesy of Gauthier and Poulin [185].*

3.4.1 Trends and Analysis

To the best of our knowledge, geometry images are applied in a small niche domain, mostly developed in academia. They are well adapted to the GPU, and therefore to real-time display, but 3D polygonal meshes remain the most common representation on the GPU. Therefore, one can suspect that geometry images will adapt well to the constant advances in GPU technology, but the difficulties in getting a good surface parameterization, central to geometry images, will limit their wider adoption.

3.5 Particle Systems

Particle systems tend to refer to the simulation of fuzzy phenomena, which are usually difficult to render with polygons. Examples using particle systems include smoke, water, trees, clouds, fog, etc. There has been older literature on directly rendering particle-based trees [475] and clouds [56, 280], but today, much of the focus on rendering particle systems has fallen on either image-based impostors (for trees, see Section 3.3), or implicit surfaces (for clouds, water, etc., see Section 3.2), or voxels (for clouds, smoke, fluids, fire, etc., see Section 3.7). Figure 3.2 illustrates the use of a particle system for a rendering of blobby-based water. The main exception of particle systems not handled by image-based impostors, blobbies, or voxels is the rendering of atmospheric shadows, also known as volumetric shadows, which is discussed in Section 5.6.

3.6 Point Clouds

Levoy and Whitted [347] introduce point clouds as a display primitive. A point cloud describes a surface using an unorganized set of 3D points (in floating-point coordinates), and each associated 3D point usually contains information such as color and normal to permit shading. If the points are sufficiently close, a surface begins to form. This representation is not unlike a polygonal mesh description, where each point is basically a polygon vertex, but a point cloud has no connectivity information between the vertices (i.e., edges and faces). See Figure 3.6 for a display of raw point-cloud data, and the insert, which is a zoomed-out view that allows all the points to be close enough to produce a smoother shading appearance.

For a while, this primitive rarely was considered in the literature. It came back into the research fold as laser and range scanners produced point-cloud data, and there was a need to model and display point clouds. Today, even common games hardware such as Kinect (for XBox 360) generates point-cloud data. Also, most computer-vision techniques to extract 3D models from photographs tend to initially generate point-cloud data, as in Figure 3.6. Examples of point cloud visualization software include (Arius) Point Stream, XB Point Stream, Resurf3D, VG4D,

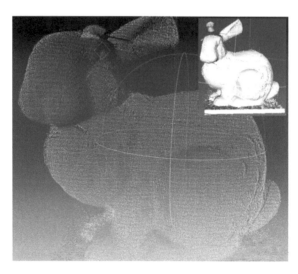

Figure 3.6. Raw point-cloud data. Notice the holes between the points; the insert shows a zoomed-out view of the same data. *Model courtesy of Chen and Yang [89].*

Cyclone, Prism 3D, etc. These software applications do not appear to handle shadows (proper self-shadowing can be difficult as well) in general, although Point Stream fakes some shadows onto a planar floor.

Fast algorithms for shadows from point clouds can be difficult because there is usually no connectivity information between the points, and there can be empty space between the points (that should be filled—see example of the empty space from Figure 3.6). While these empty regions can be filled up by some method in viewing space, this same filling may not provide sufficient filling in light space (i.e., there may be see-through problems in the shadowing); or if a different filling is used for light space, then there could be a discrepancy between the view- and light-space geometry, possibly causing bad self-shadowing problems. This is one of the reasons why a number of papers have focused on converting point-cloud data to polygons [594, 552] for rendering purposes and also converting to voxels for fast visualization. The data from Figure 3.15 actually came out of a laser scanner with significant complexity of point-cloud data. It was converted into lower resolution voxels for fast visualization. Equivalently, LiDAR data is very point cloud–like, and representation for visualization of LiDAR data is always either in polygonal form (called TIN, triangulated irregular network), heightfields, or voxels [566, 187].

The following sections describe the papers directly addressing shadow calculations for a point cloud, which include shadow depth map techniques (Section 3.6.1), ray tracing techniques (Section 3.6.2), and a few miscellaneous approaches (Section 3.6.3).

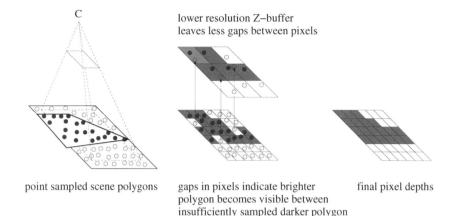

C

lower resolution Z–buffer
leaves less gaps between pixels

point sampled scene polygons   gaps in pixels indicate brighter   final pixel depths
                                     polygon becomes visible between
                                     insufficiently sampled darker polygon

Figure 3.7. A hierarchy of Z-buffers can help detect and fill gaps.

3.6.1 Shadow Depth Map

Grossman and Dally [205] use a hierarchy of Z-buffers to detect gaps. Since gaps
are found and fixed, a shadow depth map approach can be trivially used to produce gap-less shadows. See Figure 3.7 for how a lower resolution Z-buffer can help
fill holes in the higher-resolution Z-buffer. A few papers [145, 62] use variations of
splats (see description of splatting from Section 3.7.1) to gap-fill the empty space
and then easily generate the shadow depth maps. Because the above gap-filler approaches are approximate in Z-depth computation, each done in their own light or
view space, such approaches can result in bad self-shadowing. See Figure 3.8 for a
sample rendering with shadows.

Guennebaud and Gross [207] do local
surface construction by using least-squares
fitting of algebraic surfaces. As a result, the
standard shadow depth map approach is
applied without any gaps.

Gunjee et al. [212] introduce a translucent shadow mapping algorithm that uses
a spherical coordinate system, which permits omnidirectional mapping and stores
additional information per shadow map
pixel, similar to the deep shadow map [362]
(Section 5.4.2) that allows the algorithm
to handle semitransparency. Another approach is proposed by Dobrev et al. [137],
where depth peeling is used to compute
light visibility at each point on the point

Figure 3.8. Using shadow depth maps
and splats of a point-cloud model to render shadows. ©2004 IEEE. Reprinted,
with permission, from [145].

cloud, and this information is stored in a point-cloud shadow map that determines whether the pixel is lit or in shadow.

3.6.2   Ray Tracing

Schaufler and Jensen [502] use ray tracing to render and compute shadows for point clouds. A ray in this context is a closed cylinder with radius r, where r is a value slightly larger than the radius of the largest spherical neighborhood that does not contain any point. Intersection hits occur when there are points in this cylinder. The intersection hit point and normal are reconstructed as a weighted average of the nearest points inside this cylinder. The same idea is extended for shadowing, using shadow rays. Automatic determination of the r value may be difficult and requires an $O(n^2)$ preprocess step, where n is the number of points in the cloud. Structures such as octrees and k-d trees can significantly reduce this cost. Similarly, Wand and Strasser [615] ray trace cones into a multiresolution hierarchy, where the cone structure allows a weighted average of the point-cloud hits, and the multiresolution is used for intersection culling.

Instead of ray tracing points (as described above), Adamson and Alexa [1] intersect a ray against locally reconstructed surfaces via numerical iteration, which can be quite slow. Several papers [608, 351, 290] ray trace splats (disks) instead. Each of the above papers discusses different approaches to intersection culling. Wald and Seidel [608] also examine the feasibility of intersecting implicit surfaces to represent the points, but find the performance to be too slow.

3.6.3   Other Miscellaneous Techniques

An important development of point clouds comes from the concept of *surfels* [459]. Surfels are basically point samples that contain color, normal, depth, etc. Data is scanned from three orthographic views of a synthetic model and stored in hierarchical form to handle LOD. High-quality rendering is achieved through warping and texturing techniques on top of the LOD information. While this is an important development, shadowing has not been discussed. The shadowing issue may be tricky for surfels due to the changing LOD possibly causing bad self-shadowing effects if not carefully considered.

To reduce the usually large number of points to be considered for shadowing, Katz et al. [292] identify only the visible points from a specific viewpoint without reconstructing the surface or estimating normals. In this way, any shadow algorithm used can benefit from a much reduced data set to consider shadowing.

Dutré et al. [150] use a clustered set of point clouds to approximate a lot of tiny polygons forming meshes and to compute approximate visibility between the clustered set of point clouds. In this way, global illumination can be achieved in a faster manner without having to consider many tiny polygons. Because of the global illumination nature of the algorithm and of the visibility determination, this technique can produce shadows as well.

3.6.4   Trends and Analysis

From the previous descriptions, we can see that there are basically some shadow depth map and ray tracing variations to render point clouds directly. In addition, due to lack of silhouette information, shadow volume approaches have not been used for point clouds. While there are open standard, real-time libraries such as XB Point Stream and PCL that render point clouds directly, much of the industry software seems to revolve around point-cloud rendering by converting to polygons (or NURBS) or to voxels when shadows are required, as those primitives offer better speed and capabilities in terms of rendering.

3.7   Voxels

In this section, we refer to voxels as a geometric primitive [293], not voxels as supplementary data structures used in ray tracing or global illumination acceleration techniques. Voxels used in this manner are also often referred to as volume graphics. One interesting difference between the two voxel usages is that a volume graphics voxel is almost always cubical, whereas acceleration voxels can be rectangular. Also, the volume graphics voxel resolution is usually much higher than the acceleration voxels.

A voxel is basically a 3D cubic box, and a 3D grid of voxels is built to contain a larger set of voxels, which is also a 3D box (note there are non-box-like irregular grids used to deal with volumes; however, there have been no specific shadow algorithms discussed for those irregular grids to date, so we will not discuss it). Surfaces are described by assigning color (either RGB or RGBA) and normal (among other attributes) to each voxel, to permit shading. See Figure 3.9 as many voxels are used to represent an object. Like Lego, with high enough resolution (and rendering techniques), the combined voxels will appear as a smooth surface.

There are similarities between voxels and point clouds, in that a series of very small elements, expressed in (x, y, z) coordinates, is used to describe a surface. However, the main difference between point clouds and voxels is that voxels are organized in a 3D grid with implicit connectivity information between neighboring voxels, whereas point clouds are usually not organized and contain no connectivity information, although they are adaptive in terms of level of density. Another difference is that voxels can much more easily represent solids.

Voxels as a display primitive have been used for scientific visualization, such as viewing X-rays, CT-scans, MRIs, LiDAR data, point clouds, or engineering parts (e.g., polygons [113]). These scientific visualization examples are also sometimes simply referred to as volume graphics. In the above examples, voxels are usually not natively created, but other sources of data are converted into voxels, and they are usually static objects.

Volume graphics examples that are dynamic in nature include terrains, geological activities, seismic activities, computational fluid dynamics, abstract

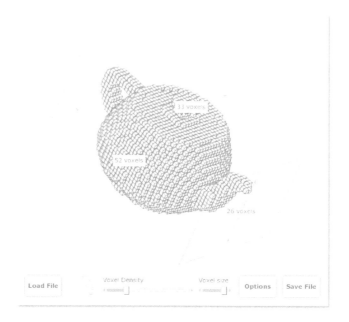

Figure 3.9. Voxels as indicated by the boxes within a voxel set (grid). *Image courtesy of Arjan Westerdiep.*

mathematics, weather, etc. Voxels have also been used more and more in film special effects for clouds, smoke, fire, etc., with software from JIG, Hypervoxels, etc. Note that voxels are typically semitransparent (as encoded in the A of the voxel's RGBA). This is another potential difference with most point-cloud data, where point clouds are usually assumed to be opaque.

In this section, the coverage on shadows from voxels will include several classes of algorithms, with all of the research papers focusing on uniform grids. These classes of algorithms are all meant for real-time applications, except for the CPU-based ray tracing approaches:

- Splatting (see Section 3.7.1).

- Shear-warp (see Section 3.7.2).

- Slice-based hardware textures (see Section 3.7.3).

- Ray tracing techniques (see Section 3.7.4).

- Shadow depth map techniques (see Section 3.7.5).

Technology has advanced significantly in this domain in recent years. Splatting, shear-warp, and slice-based hardware textures, although prominent earlier, are less used today due to the emergence of GPU-based ray tracing and shadow depth map techniques, which can achieve higher quality, good performance, and other capabilities (e.g., motion blur, multiple voxel sets, etc.).

3.7.1 Splatting

The splatting approach [628] does voxel-order traversal in a front-to-back order from the perspective of the camera view and splats the occupied voxels by projecting an appropriately sized sphere onto the screen (see Figure 3.10).

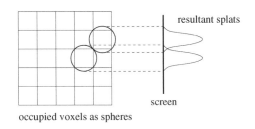

Figure 3.10. Splatting technique by projecting a splat onto the screen.

In terms of shadow generation from splatting, Nulkar and Mueller [436] use an image-aligned sheet-buffer variation of splatting [419] to achieve shadows by accumulating opacity values from the light-source rendering into the voxel data, but this technique requires an entire 3D voxel set to store opacity values accumulated in the direction of the light. During camera rendering, this opacity value is accessed to determine the amount of shadowing (see Figure 3.11). Grant [202] has done something similar by only requiring a 2D shadow buffer instead of a 3D voxel set. However, there are artifacts for the above techniques when the viewing and lighting directions are too different.

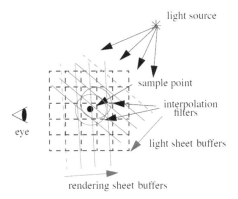

Figure 3.11. Splatting shadows using image-aligned sheet buffers. *Image courtesy of Nulkar and Mueller [436], ©Eurographics Association 2001. Reproduced by permission of the Eurographics Association.*

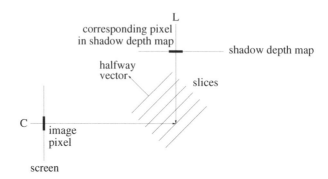

Figure 3.12. Using the halfway vector as the orientation for the slices to be splatted.

Two papers [657, 310] improve upon both approaches by incrementally accumulating the opacity values for shadow determination by storing the values in
a 2D shadow buffer. To get a consistent orientation of the image-aligned sheet
buffer between the light and the camera, the halfway vector H is used as the orientation to suit both the camera and the light. This common orientation between
the camera-view and light-view slices is needed, or the order of traversal between
light and view is different and cannot rely on only a 2D shadow buffer (see Figure 3.12). Likely, problems will occur when the camera and light views are too far
apart.

Zhang et al. [658, 662] extend their approach by storing participating media
information in the 2D shadow buffer. As well, texture-convolution is applied on
the 2D shadow buffer to achieve soft shadows. This combination does not contain
the disadvantages of original texture convolution approaches because the occluder
and receiver are axis-aligned slices (in the 2D shadow buffer), so the need for approximate occluder and receiver planes is not necessary.

3.7.2   Shear-Warp

Lacroute and Levoy [322] introduce the shear-warp factorization to achieve fast
front-to-back voxel rendering. The voxel set is warped so that screen axis-aligned
renderings can be done (see Figure 3.13), and a clever approach to skip already
occluded voxels is introduced. Shadows are not considered in this work except in
Lacroute's dissertation [323], where shadow generation is attempted by extending
Grant's approach [202], but the authors continue to have the same quality problems
that Grant experienced. If the H halfway vector extension [657, 310] is applied, it
may help resolve those problems.

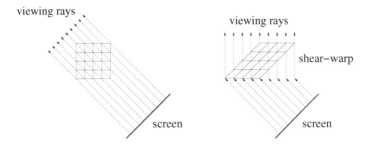

Figure 3.13. Shear-warp technique to allow fast axis-aligned rendering.

3.7.3   Slice-Based Hardware Textures

Behrens and Ratering [44] compute shadows in a hardware texture-based volume-rendering approach introduced by Cabral et al. [78]. In this case, n slices of accumulated shadow information are computed, where each slice indicates a slice in the voxel representation. The accumulation of shadow information is done by projecting the shadow information from a particular slice that is closer to the light to all slices farther away from the light. This is repeated for each slice in an incremental fashion so that the algorithm runs in linear time. The shadow information is then blended on the GPU into the illumination of the voxels. Due to this accumulation, the shadowed results tend to be overblurred and sometimes darker than they should be.

Ikits et al. [255] extend the technique from Zhang and Crawfis [657] by using GPU fragment shaders to blend the alpha values with a 2D shadow buffer. While this approach provides improvements to the Behrens and Ratering [44] approach, there remain cases where the shadows are overblurred and sometimes darker.

3.7.4   Ray Tracing

CPU Ray Tracing

Levoy [348] applies ray tracing with a fast voxel traversal scheme. A voxel in this paper (and most ray tracers) indicates an enclosure of eight points at the corners of the voxel versus a single point in the middle of the voxel, which is the case for the papers discussed thus far. This enclosure of eight points at the corners of the voxel is sometimes referred to as a cell. When a ray hits a voxel with any of the eight points non-empty, then there is a hit, and the resultant hit and its illumination is a trilinear interpolation of those eight points. Shadow rays are treated in the same manner.

Jones [277] extends the above approach with some precomputed occlusion values. During preprocessing, the voxels are computed whether they are inside a solid region or on the surface (labeled "transverse voxel"). Rays are shot to the light

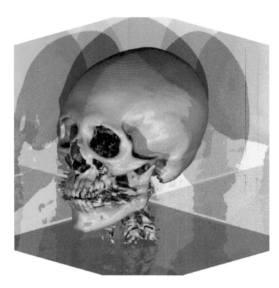

Figure 3.14.   Ray tracing voxels to achieve shadows.   *Image courtesy of Jones [277],
©Eurographics Association 1997. Reproduced by permission of the Eurographics Association.*

source, and when the traversal scheme hits a voxel that indicates it is already inside a solid region, it is in shadow. If not, then traversal checks whether it hits a
transverse voxel. If traversal through the transverse voxels comes up empty, then
it is fully lit. Otherwise, the actual surface is intersected against the shadow ray
to determine shadow occlusion. With the above approach, much less ray-surface
intersections should need to be done. See Figure 3.14 for a sample rendering.

Discrete ray tracing [650] uses ray tracing of the opaque voxels to determine
shadow occlusion. The shadow occlusion is encoded as a bit in each voxel, whether
a voxel is in shadow or not. In a static environment, querying this bit indicates
whether the region inside this voxel is in shadow or not.

GPU Ray Tracing

Sobierasjski and Avila [549] use a hardware render pass to obtain nearest and farthest depth per pixel and use this information to accelerate the actual ray tracing
computations. With the information of nearest depth per pixel (actually the nearest depth of its own pixel as well as neighboring pixels), the ray tracing for that
pixel can start at that depth. This means that much ray traversal computation is
avoided. A quick exit condition is achieved when the ray has traversed beyond the
farthest depth. This is repeated for the shadow pass for simple light sources such
as infinite, point, or spot lights. For totally opaque surfaces, only the nearest depth
is of concern. For shadows from semitransparent voxels, the farthest depth can be
useful. With this approach, however, small details can be missed if not caught by

the hardware render pass because the hardware render pass may not capture potentially important subpixel regions closer to the camera view that the ray tracer would have caught otherwise (without this optimization step).

Ray tracing on the GPU has become feasible due to the availability of advanced fragment shaders [488, 319]. Many of the techniques used for GPU ray tracing have been adopted from CPU techniques relating to voxel traversal [198, 526, 160] with a large focus on ray tracing acceleration relating to early ray termination, fast voxel traversal, and efficient space leaping, usually applying some variants of sparse octrees [490, 70, 112, 113, 491, 326]. Shadow rays are then a natural extension for shadow computations on the GPU.

Note that in the context of volume graphics, since voxels are already a part of the data structure, voxel-like structures (such as sparse octrees) are often used for performance reasons instead of the application of k-d trees, BSP trees, or bounding volume hierarchies as seen in polygonal intersection cullers (Section 2.5.2).

3.7.5   Shadow Depth Map

A variation of the shadow depth map approach can be used if all voxels are either empty or opaque [490], where a pregenerated shadow depth map is rendered, and where the distance from the light to the closest voxel is stored in the shadow depth map. When rendering a voxel, it is projected onto the shadow depth map, and if

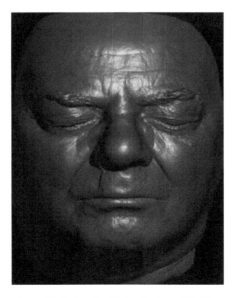

Figure 3.15. Voxel-based model of a head with its shadows. *Model courtesy of XYZ RGB Inc. and image courtesy of NGRAIN (Canada) Corporation.*

the distance of the voxel is farther away from the shadow depth map value, it is in shadow; otherwise, it is not (see Figure 3.15 for a sample rendering). To avoid self-shadowing (Section 2.3.4), the bias approach is chosen, as the surfaceID or mid-distance approaches are not easily extended for voxels. However, the bias seems a good offset in the direction of the voxel normal.

If there are semitransparent voxels, then the GPU-assisted variation of the deep shadow map [217, 490] can be applied in the same manner—see the description of the deep shadow map algorithm in Section 5.4.2. The GPU-assisted variation for voxels uses GPU ray casting to do initialization of the deep shadow map information because the ray tracing operation implicitly generates a sorted list of depth values without the expense of sorting.

3.7.6 Trends and Analysis

- Fourier volume rendering [163] is another approach to directly render voxels. However, shadowing has not been attempted with this approach and thus is not included in the above discussions.

- Conversion to polygons [367] remains a very popular option in the industry, and it makes sense because there are many rendering options available for dealing with polygons. However, with recent advancements, rendering natively in voxels has become very practical.

- While there are techniques to store shadow volumes as occlusion bits in voxels [650], there has been no application of the shadow volume approach in this domain.

- If all voxels are entirely either empty or opaque (i.e., each voxel is only RGB), then the simple shadow depth map can be very efficient and effective. For voxels that contain semitransparency, GPU-based deep shadow maps can be computed in real time, although ray tracing still provides the highest quality, and its usage on the GPU has achieved much faster speeds.

- In the visualization industry, NGRAIN uses the basic shadow depth map approach for fast shadowing, VolumePro uses the combination of ray tracing and shear-warp on a customized chip, VTK uses GPU ray tracing, and VGL ray traces isosurfaces. For film special effects, deep shadow maps [645] (as used in production) or ray tracing (as used in Hypervoxels) are the norm.

- There is an increasing number of scenarios where multiple voxel sets need to be rendered [70], accounting for correct shadows from possibly intersecting voxel sets. In the extreme case, NGRAIN helps with maintenance and repair of complex mechanical equipment (e.g., jet engine), where each mechanical part is a voxel set. There can be thousands of such voxel sets because each mechanical part may need to be assembled or disassembled properly. For most ray tracing [70] or shadow depth map approaches, shadow computation is generally not a problem. However, extensions need to be developed for shadow buffer–based shadow solutions [202, 436, 657] and slice-based hardware texture solutions [44, 255]. This is a problem because the multiple voxel sets can intersect (although the geometry itself does not intersect), and the semitransparent shadowing function needs to be composited correctly in the depth order of the semitransparent regions, which may jump from one voxel set to another.

- The need for handling large voxel-based data sets is growing. Although this is not specific to shadowing within voxels, voxel shadow algorithms do need to integrate with solutions for large voxel data sets. LOD and out-of-core (e.g., bricking) techniques are available in [113, 258] and Chapter 17 of [160]. Also note that most out-of-core techniques favor ray tracing approaches due to the ease of swapping in and out of chunks (bricks) of voxels at a time.

3.8 Heightfields

There are some references on the web and in the industry where the term voxels is misused to indicate heightfields. Although voxels can be enumerated to represent heightfields, voxels are 3D elements, and can handle concavities that heightfields cannot. A heightfield is a 2D uniform grid, where each grid element is associated with a single height value (relative to a flat surface) to generate common geometries such as terrains or mountains. See Figure 3.16 for raw data after texture mapping to a polygon and the rendering result. The source of the raw data can come from point clouds, LiDAR (although multireturn LiDAR data cannot be handled

Figure 3.16. The heightfield data when texture-mapped to a polygon and its rendering result. *Image courtesy of Ratko Jagodic.*

by heightfields, but can by voxels), topographical maps, USGS data, etc. Height-fields are also sometimes referred to as height maps.

Note that heightfields are quite similar to bump or displacement maps (Section 5.2). The differences are that heightfields assume a planar base (while displacement maps can be applied on a curved surface), and heightfield shadowing needs to be a lot more precise than for a displacement map since the camera will often lie very close to the heightfield. This similarity leads to a discussion about sharing similar shadow algorithms (see Section 3.8.2).

Earlier research involves keeping the shadow information provided by projecting the scene in the light source direction [409, 485]. Since then, several direct and indirect rendering approaches have become feasible options.

In terms of indirect rendering approaches, heightfields can be converted into polygons [350] (including handling LODs) so that the GPU can easily be used for rendering needs, including shadows, of which there are many algorithms to choose from within this book (e.g., Kaneda et al. [289] apply shadow volumes). As alluded to in the previous paragraph, there is also opportunity for mapping heightfields to voxels [105], and the voxel rendering options made available (Section 3.7).

In terms of direct rendering options, variations of ray tracing (Section 3.8.1) and horizon mapping (Section 3.8.2) have become the preferred approaches, the former due to their simplicity, the latter due to their ability to utilize the GPU for real-time performance. However, care must be taken with horizon-mapping approaches due to special casing when heightfields need to integrate with other geometry types. More are discussed in terms of the direct rendering options below.

3.8.1   Ray Tracing

Using ray tracing, the 2D uniform grid is traversed, using any DDA algorithm, from the point to be shaded to the light source. If the visited pixels from the 2D grid indicate that the pixel height is higher than the ray height, then that point is in shadow [111]. A GPU-assisted version of the above is also available [381]. However, this basic approach results in the rendering of flat tops, which means the highest points are always horizontally flat. To achieve pointy tops, Musgrave [420] bilinearly interpolates the height of the four corner points. To achieve proper ray hits, he uses the above traversal approach, but if the minimum height of the ray extent (crossing the pixel) is higher than the maximum height of the four corners, then traversal continues assuming no hit; otherwise, the ray is intersected against two triangles formed with the four corners to determine ray-hit success.

Performance gains in ray tracing heightfields can be seen in a couple of directions, including the work of Henning and Sephenson [239], that optimizes the traversal algorithm to achieve faster rendering speeds, and the work of Qu et al. [471] that uses the GPU to render the triangles.

Figure 3.17. Heightfields used for visualizing terrain using a hierarchical approach. ©*1998 IEEE. Reprinted, with permission, from [562].*

3.8.2   Horizon Mapping

As for horizon mapping, as stated in Section 5.2, the shadows are approximate. While appropriate for bump maps, this is not appropriate for heightfields due to the high frequency changes that can be abundant in heightfields. As a result, sampling in more directions is necessary, and thus the research has focused on faster computation of the horizon angles.

Stewart [562] extends the horizon-mapping technique by using a hierarchical approach to consider all points in the heightfield as occluders. See a rendering result in Figure 3.17. Snyder et al. [547] apply a multiresolution pyramid of height

Figure 3.18. Heightfields used for visualizing mountains, using a multiresolution pyramid of height values. *Image courtesy of Snyder et al. [547]. Computer Graphics Forum ©2008 The Eurographics Association and Blackwell Publishing Ltd. Published by Blackwell Publishing, 9600 Garsington Road, Oxford OX4 2DQ, UK and 350 Main Street, Malden, MA 02148, USA. Reproduced by permission of the Eurographics Association and Blackwell Publishing Ltd.*

values—at a given point, the check for maximum height values uses high resolutions for close-by grid pixels and lower resolutions for far-away grid pixels. Thus, the details close by are not lost, and the far-away details are blurred to generate soft shadows, so extreme accuracy is not as crucial. See a rendering result in Figure 3.18. Timonen and Westerholm [586] also accelerate the computation of the horizon angles, by being able to reuse neighboring grid pixels' height and horizon angles as guide for the current pixel's horizon angle.

3.8.3   Trends and Analysis

Most of the industry software do not apply direct rendering of heightfields when heightfields are not the only primitive to render and using direct rendering techniques would incur nontrivial amounts of processing in order to integrate the different primitives. Example integration issues include heightfield shadows on other objects and other objects' shadows on the heightfield. As a result, when rendering heightfields, industry software such as ArcGIS, Terragen, Picogen, POV Ray, Blueberry3D, etc., convert to polygons for rendering needs, or the native format is already in TIN (triangulated irregular network) format. There are also some growing trends to convert heightfields to voxels as well, such as the engines from Vulcan 3D and Atomontage. Both polygons and voxels offer good integration and rendering capabilities.

3.9   Final Words

While polygons remain the dominant representation, and polygonization is often a very feasible approach, rendering in its native form remains an important research area to pursue. This is especially true for representations that present some advantages that polygons do not possess, for example, where voxels can represent solid information at a granularity that polygons cannot, or rendering of particle systems that can be more efficient using image-based impostors or implicit surfaces.

Next up, soft shadows!

CHAPTER 4
Soft Shadows

4.1 What's Covered in This Chapter

In addition to the visually pleasing aspects of soft shadows, studies by Rademacher et al. [472] indicate that the presence of soft shadows contribute greatly to the realism of an image, which makes the generation of soft shadows that much more important. In this chapter, we only discuss soft shadows from extended light sources, such as linear lights, polygonal, spherical, and general area lights, and volumetric lights. Several excellent surveys [227, 7, 158] exist on this topic. Other sources of soft shadows are discussed in Chapter 5; they include motion blur (Section 5.7), ambient occlusion (Section 5.8), precomputed radiance transfer (Section 5.9), and global illumination (Section 5.10).

We start off this chapter with some soft shadow basics (Section 4.2), followed by some theory behind soft shadow computations (Section 4.3)—the latter is very important to understand though the concepts have not received much adoption to date. The main categorized approaches are then presented, in the same order as they were presented for hard shadows, including the planar receivers of shadows algorithms (Section 4.4), shadow depth map (Section 4.5), shadow volumes (Section 4.6), ray tracing (Section 4.7), and some miscellaneous additional soft shadow algorithms (Section 4.8). The algorithms introduced for soft shadows can vary in terms of accuracy, which can affect the effectiveness of certain algorithms on certain applications. However, the goal of all soft shadow approaches is the aesthetics of the rendered results and performance feasibility, which are discussed in this chapter. The chapter concludes with trends and analysis of the criteria of applications and appropriate deployment of each categorized approach (Section 4.9).

4.2   Soft Shadow Basics

4.2.1   Computing the Soft Shadow Integral

In Section 1.2.2, recall that a fraction in the range of [0,1] is multiplied with the light intensity, where 0 indicates umbra, 1 indicates fully lit, and all values in between indicate penumbra. This statement is actually incorrect, and thus it is important to understand the nature of the integral that must be evaluated in order to compute the soft shadow resulting from the direct illumination of an extended light source. The correct integral of the irradiance E at a receiving (illuminated) surface element dA can be expressed as follows:

$$E = \int_{A_e} V \left(\frac{L \cos \theta \cos \phi}{\pi r^2} \right) dA_e,$$

where V is the binary visibility of an emitting surface element dA_e on the light source, and the remainder of the equation is the irradiance over a surface element as if fully lit.

   Ideally, if the domain of integration can be reduced to the fragments of the extended light that are visible from the point to be shaded (i.e., the points on A_e for which V is 1), then the integral reduces to a direct illumination integral over precisely these fragments. Unfortunately, as will be seen in back-projection algorithms [141, 561], determining these fragments is a difficult problem in itself, as visibility must be evaluated for each fragment dA_e for any visible 3D point. Thus a common simplification for the computation of the integral is to assume that it is separable, and to integrate visibility and irradiance separately. Many of the soft shadow algorithms presented in Chapter 4 will assume this separation of the visibility term. This gives rise to the following approximation of the irradiance:

$$E = \left[\int_{A_e} V dA_e \right] \left[\int_{A_e} \frac{L \cos \theta \cos \phi}{\pi r^2} dA_e \right].$$

   Note that whereas this approximation will generate soft shadows, the accuracy of the results will not necessarily be reliable, especially in cases where the solid angle subtended by the light source is large (e.g., the light source is large and/or close to the receiving surface) or the extended light's normal is close to perpendicular to the surface's normal. The inaccuracies from the decoupling of the irradiance integrand and of the visibility factor will be most evident when the parameters θ, ϕ, and r vary greatly over the domain of the integrand and the visibility function V is not constant. For example, in Figures 4.1 and 4.2, the same fraction of the linear light's length is occluded from the point to be shaded. However, whether the shadowing function is evaluated outside or inside the irradiance integral leads to different results, as the solid angle formed by the visible parts of the light is quite different.

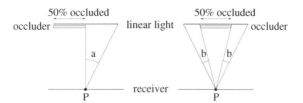

Figure 4.1. The same occlusion fraction of an extended light source may not result in the same illumination value.

The appeal of such a simplification is great, given that an analytic solution can be computed for the non-shadowed illumination integral [430, 465, 578] for specific reflection models and that some algorithm can be used to approximate the visibility of the light (i.e., the integral of V over A_e). The approximation is also simpler for radiosity solutions (see Section 5.10.1), for some direct illumination algorithms [550, 669], for sampling criteria in the final gathering of penumbrae [504], and may also be applied to specular integrands. However, care must be taken that the artifacts resulting from such an approach are reasonable. This being said, such approximations are more than acceptable in several visually plausible situations. In fact, the differences are not visually discernable in many situations and applications, thus explaining their popularity.

Another approximation consists of computing individual shadows and then combining them. This remains an approximation, as the combined occlusion (also

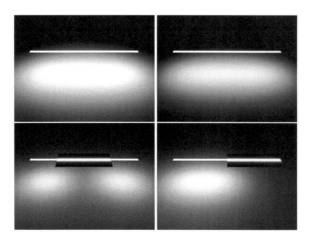

Figure 4.2. In the top row: shading without shadowing (left), and shading with a constant shadowing fraction (50%) for all pixels (right). In the bottom row: shading with correct pixel-wise shadowing fraction (50%, but within the integral) from an occluder at the center of the light source (left), and to the right side of the light source (right).

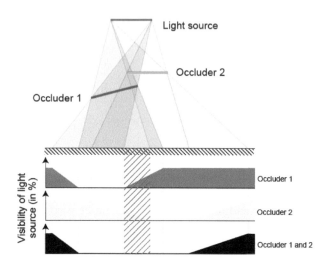

Figure 4.3. The shadow cast by a set of objects can be quite different than the combined shadows of each individual object. *Image courtesy of Hasenfratz et al. [227], ©Eurographics Association 2003. Reproduced by permission of the Eurographics Association.*

called occluder fusion) can be quite different than summing up occlusion factors for each individual shadow. This is illustrated for a simple case in Figure 4.3 and for an actual 3D rendering in Figure 4.4.

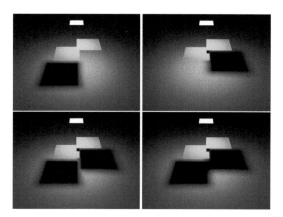

Figure 4.4. In the top row: correct shadow cast by the left occluder only (left), and by the right occluder only (right). In the bottom row: combined shadows as the maximum occlusion of the two individual shadows (left), and correct shadows from both occluders (right).

4.2.2 Generic Algorithmic Approaches

From an algorithmic standpoint, the most straightforward approach to consider soft shadow generation from extended lights is to apply any existing hard shadow algorithm for a point light source and simulate the extended light shadow with many point lights, where each point light represents a small region of the extended light. This is actually the basis for the approach by Haeberli [218], Brotman and Badler [71] (Section 4.5.4), for distribution ray tracing [108] (Section 4.7.1), and Herf and Heckbert's [240] approach (Section 4.4.1), among others. However, just a straightforward implementation of multiple point source simulation tends to be very slow, often requiring a large amount of memory for the shadowing structures, and usually results in separation or banded results, i.e., there are noticeable shadow boundaries from the point sources (see Figure 4.5 as an illustration of this separation/banding artifact (left) versus smooth soft shadows (right)). This produces a very low-quality result because our human visual system is very sensitive to and unforgiving of banding artifacts. Banding from multiple point sources is reduced if

- Many such point sources are used, e.g., Herf and Heckbert [240] use as many as 256–1024 point sources to simulate an extended light, which results in slow rendering times. The difficult question becomes then how many point sources are sufficient, and this question is barely addressed at all in the literature.

- Some low-discrepancy, nonuniform, or stochastic sampling patterns are employed, as in distribution ray tracing techniques (see Section 4.7.1). Such patterns usually result in noise (to mask the banding artifacts), and our human visual system is a lot more accepting of noise than banding.

Figure 4.5. Treating soft shadows as an averaging of hard shadows from multiple point sources can result in shadow bands (left), but ultimately converges to proper soft shadows (right). *Image courtesy of Hasenfratz et al. [227], ©Eurographics Association 2003. Reproduced by permission of the Eurographics Association.*

- Some filtering, interpolation, or blending is done between the multiple samples.

Due to the above performance and banding issues, multiple point sources have not been the only approach explored in the literature. Many of the algorithms that are discussed in this chapter tend to start off from a hard shadow preprocess, then perform operations to approximate soft shadows. This can be seen in plateau shadows (Section 4.4.2), z-difference penumbra approximation (Section 4.5.1), micropatches (Section 4.5.2), silhouette detection-based solutions (Section 4.5.3), penumbra wedge implementations (Section 4.6.3), single-ray analytic solution (Section 4.7.3), and ray tracing depth images (Section 4.7.4). Because the results are approximated from a hard shadow standpoint, the results are not physically correct but can be visually pleasing to different degrees.

For further performance reasons, the concepts of an inner penumbra and an outer penumbra are introduced (see Figure 4.6). The inner penumbra is the soft shadow region inside the umbra produced from a single representative point source. Algorithms computing just the inner penumbra usually result in undersized shadows. The outer penumbra is the soft shadow region outside this umbra, and algorithms computing just the outer penumbra usually result in oversized shadows. This distinction is useful in some approximations because certain algorithms compute for performance reasons only the outer penumbra [452, 220, 83, 646] or just the inner penumbra [254, 309]. Physically accurate results should produce both inner and outer penumbrae.

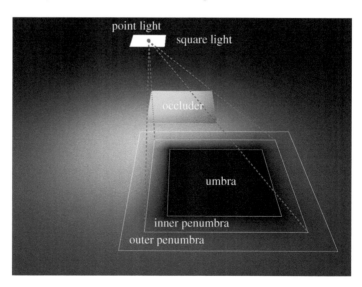

Figure 4.6. Inner and outer penumbrae.

4.3   Theoretical Aspects

Shadows result from complete (umbra) and partial (penumbra) occlusion of a light source. The occlusion function, i.e., the parts of a light source that are occluded (or equivalently, "seen") from any 3D point in space, changes depending on how objects occlude the light source. The projection of occluders from any 3D point to be shaded onto the light source, changes qualitatively at certain geometrical configurations of light and occluders. Knowing what occludes light and how it occludes it can lead to very accurate shadows.

Computing the different 3D shadow regions cast by one or more blockers for a given light source can be summarized as an interesting geometrical problem. At first, this problem could appear simple, considering that computer graphics has developed several representations and algorithms to handle some related situations, for instance, with CSG operations and BSP representations. Moreover, a knowledge of these regions can also be used for more efficient sampling strategies in order to better approximate shadowing. However, Figures 1.7, 1.8, and 4.4 barely hint at the difficulties. In reality, robustness issues, algorithmic complexity for large scenes, and memory restrictions in the number of resulting polygons, indicate some of the intricacies of the problem. This section covers these issues.

First, we restrict the theoretical aspects discussed here to polygonal scenes. This encompasses planar polygonal light sources and objects. Point and directional light sources are degenerate cases that are already handled by hard shadows computed with shadow volume techniques, as discussed in Section 2.4. There are some rare exceptions to polygonal scenes, for example, simple primitives such as spheres in cone tracing [10] and disks in a number of light-source representations for shadow mapping (e.g., [550]), but those remain more limited in application.

It is important also to note that in the history of the main contributions of this visibility theory to computer graphics, most contributions originated from the radiosity literature, which we touch on in Section 5.10. Therefore, several shadowing algorithms predating 1990 will not discuss computing shadow regions in the light of the theoretical aspects presented in this section.

One direct application of this theory for shadowing aims at splitting scene polygons into fully lit by a given light source, fully shadowed (umbra), and partially lit/shadowed (penumbra) polygons. The polygons in penumbra can "see" different parts of the light source resulting from different occluder parts, and therefore, each penumbra region is qualified by the ordered list of edges and vertices from the blocking scene polygons and light source.

These split polygons result in view-independent meshes, and in the particular case of diffuse surfaces, their shading can be precomputed as an approximation in the form of a texture or an interpolation from the color at each vertex of the polygons.

4.3.1 Discontinuity Events

The PhD dissertation of Durand [149] covers some of the more theoretical concepts in this section.

Visual events result in changes in visibility of the light source, and they occur at combinations of scene and light-source edges and vertices. Unoccluded edge-vertex (denoted EV) events form a plane such that on either side, a convex light source starts to appear or disappear. A combination of three unoccluded edges (denoted EEE) may form a ruled quadric surface with the same properties. These

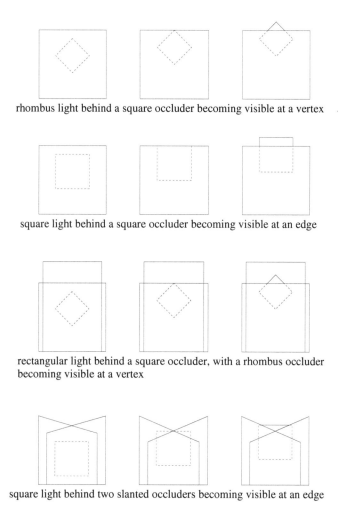

rhombus light behind a square occluder becoming visible at a vertex

square light behind a square occluder becoming visible at an edge

rectangular light behind a square occluder, with a rhombus occluder becoming visible at a vertex

square light behind two slanted occluders becoming visible at an edge

Figure 4.7. Configurations of edges leading to EV and EEE visual events.

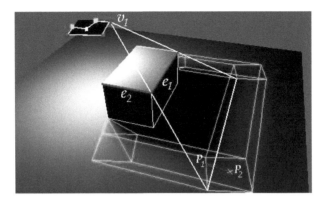

Figure 4.8. Back-projection of associated events onto the light source in order to evaluate its visibility. ©1994 ACM, Inc. Included here by permission [141].

events are illustrated in Figure 4.7. Similar events can also be critical when none of the edges and vertices belong to the light source, but on the condition that the associated visual event extends to intersect the light source. All these events, if they are not blocked by other polygons, define the various regions (also known as *aspects* of the light source) in 3D space. For any point within such a region, the back-projection through this point of all its associated events form the same ordered list of elements defining the visible portion of the light source (Figure 4.8).

Discontinuity events result in discontinuities in the shading of a diffuse polygon at the specific locations of visual events. They have been categorized according to the type of discontinuity in the shading: D^0 for shading value, for instance, where a polygon intersects another polygon; D^1 for shading derivative, for instance, between umbra and penumbra regions; and D^2 for the shading second derivative, for instance, within different penumbra regions. Such events are illustrated in Figures 4.9 and 4.10.

Nishita and Nakamae [430] compute one shadow volume for a convex occluder for each vertex of a convex light source. The convex hull of all shadow volumes form the penumbra contour, according to the authors, while the intersection of all shadow volumes form the umbra contour. In doing so, they neglect several visual events discussed above. Campbell and Fussell [80] compute also the contour of the penumbra regions and intersection of the umbra regions with minimum and maximum extremal planes. The corresponding volumes are merged with BSP representations similar to Chin and Feiner [92]. While they observe different regions in the penumbra region, they still miss some regions.

Heckbert [234] and Lischinski et al. [353] detect and compute D^0 and D^1 events in order to mesh scene polygons for an improved patch definition in radiosity algorithms. The events are projected as line segments onto the receiving polygon, and a structure is designed to connect these events or erase those that are occluded.

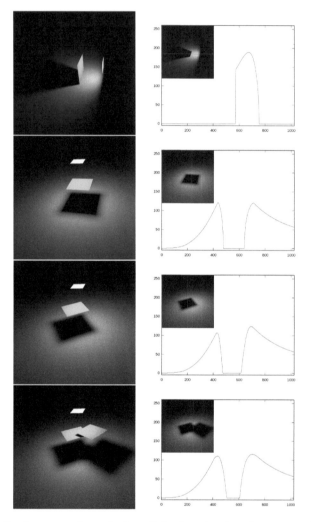

Figure 4.9. Simple scenes of cast shadows (left column). Plot of intensities along the red segment indicated in the only-shadows version of the image (right column).

Lischinski et al. [353] also use a BSP algorithm to perform these operations. They both mention D^2 events, producing quadric ruled surfaces as a future extension to their work. Drettakis and Fiume [141] and Stewart and Ghali [561] use back-projection onto the light polygon to compute a more accurate irradiance on diffuse polygons.

While these events are complex and difficult to compute robustly, Durand et al. [147] exploit the concept of changes along these events in the form of

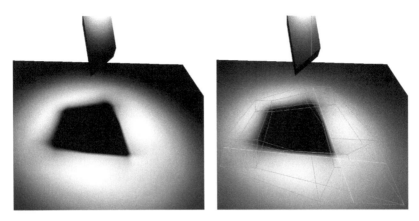

Figure 4.10. Different regions in the penumbra (left). Same configuration but with the discontinuity events indicated by a color mesh on the floor (right). *Image courtesy of James Stewart.*

extremal stabbing lines [584] defined as VV, VEE, and EEEE to incrementally compute a *visibility skeleton*. This structure corresponds to a graph where a node is an extremal stabbing line, and a link between two nodes indicates that a visibility event (a surface defined by a line swath between two extremal stabbing lines) occurs between them. With a careful implementation and catalogs of events, Durand et al. are able to handle discontinuity events for scenes of up to a few thousands polygons.

Duguet and Drettakis [144] apply interval arithmetic to the construction of these extremal stabbing lines to achieve a fairly robust implementation of discontinuity events in scenes of a few hundred thousands polygons.

Even though the maximum number of visual events can be in $O(n^4)$ for n edges in a 3D scene, this complexity can be strongly reduced considering the special case of one polygonal light source and all combinations of EEEE that do not result in an extremal stabbing line due to its construction or occlusions. Unfortunately, even given these advances, the complexity and robustness issues are still such that even today, the applications of shadowing that rely on these more theoretical bases are rare. Such examples include [418]. They remain, however, important in understanding the issues related to the complexity of computing exact shadows.

4.4   Planar Receivers of Shadows

In this section, we describe some soft shadow algorithms that assume simplified environments or shadows on planar shadow receivers only. They include soft shadow textures (Section 4.4.1), plateau shadows (Section 4.4.2), and convolution textures (Section 4.4.3).

4.4.1   Soft Shadow Textures

Herf and Heckbert [240, 232] take advantage of fast hardware Z-buffer and texturing to stochastically sample an area light source at many points. For each such sample point (on the light), and for each polygon that acts as an occluder, the shadow is projected onto all other planar receivers to form umbra regions stored as attenuation shadow maps. Essentially, each sample point pass produces hard shadows. In the end, all the results are averaged via an accumulation buffer to get penumbra regions. If enough sample points are used (e.g., 256 samples), the results can look quite good and very accurate. However, for real-time purposes, if a small number of sample points is chosen, then banding (between the discrete sample points) results. To improve performance, Isard et al. [259] distribute the above attenuation shadow maps as hardware textures, over texture units on each graphics card, and claim fast performance numbers.

4.4.2   Plateau Shadows

Haines [220] uses the basic approach from Parker et al. [452] (Section 4.7.3) to achieve soft shadows in a real-time environment but assumes that the receiver is always a planar surface. The umbra region is always the same as if it came from a point source. As in Figure 4.11, a conical region and sheet-like region are created outside the umbra silhouette vertices to define the penumbra region. The penumbra region is a function of the distance from the occluder and the distance from the umbra vertices. The algorithm can be implemented on the GPU by using the Z-buffer to create the soft shadow hardware texture, which is then mapped onto the receiving surface. Both approaches assume the computation of only the

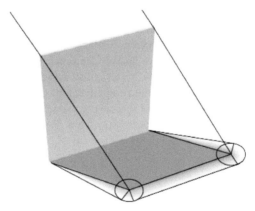

Figure 4.11. In the plateau approach, a vertex generates a cone, and an edge generates a sheet. Hardware texturing interpolates the soft shadowing. *Image courtesy of Haines [220], reprinted by permission of Taylor and Francis Ltd.*

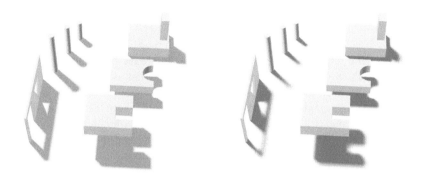

Figure 4.12. Visual comparison between the plateau algorithm (left) and the soft shadow textures (right). *Image courtesy of Haines [220], reprinted by permission of Taylor and Francis Ltd.*

outer penumbra and can produce overstated shadows. However, the results tend to look quite good (see Figure 4.12 for a comparison between the approaches of Haines [220] and Herf and Heckbert [240, 232]).

4.4.3   Convolution Textures

Soler and Sillion [550] realize that the shape of the penumbra is a function of both the occluder and the light source shapes. In an ideal situation, where the light source, occluder, and receiver are parallel to each other, this function corresponds to a convolution of the occluder and the light source. This is achieved by rendering the occluder and light source into shadow maps without Z-buffering turned on. In other words, convolution, in its simplest form, means that the shadow of the occluder is blurred, with the distance from the occluded object controlling the amount of blur. They choose some heuristics as representative and virtual light sources, occluders, and receivers projected onto parallel planes. However, choices of such virtual representations can result in errors and can easily miss self-shadowing cases. In scenarios such as meshes for radiosity, these limitations may be less severe.

Eisemann and Décoret [156, 157] extend the above approach. To improve results from overlapping shadowed objects, slices (buffers) parallel to the polygonal light are created, containing the prefiltered occlusion textures. For each point to be shaded, the slices from the point to the light are intersected (similar to ray tracing depth images), and a convolution of the slice results is performed. While the improved method can lead to plausible shadows, it still can suffer from some problems when using too few slices and lower texture resolutions. These problems can appear as incorrect self-shadowing, light-leaking problems, complex overlap of shadowed objects, etc.

4.5 Shadow Depth Maps

There are basically five categories of shadow depth map algorithms to produce soft shadows. The first two categories include a z-difference penumbra approximation approach (Section 4.5.1), and a recent approach that uses micropatches (Section 4.5.2) and back-projection to approximate soft shadows. Both use a shadow depth map from a single point source. Furthermore, there are the silhouette-based shadow depth maps (Section 4.5.3), which rely heavily on silhouette properties and are represented by one or a few point light sources. The fourth category of shadow depth map approaches is the multiple point source approaches (Section 4.5.4), which tends to be slower though more physically accurate. The final category discusses artifacts due to low resolution when rendering large scenes (Section 4.5.5).

The first two categories of algorithms should be the fastest, but their performance can be hampered if the extended light is very large (unless prefiltering techniques are used). The z-difference penumbra approximation is the simplest to implement, least accurate due to the assumptions made, and most worry-free in terms of artifacts, whereas care must be taken to avoid artifacts when using micropatches. The silhouette-based shadow depth maps are faster than the multiple point source approaches but also generate the most artifacts, including incorrect shadows at the intersections. The multiple point source approaches are the most expensive if many point sources are needed, but noise or banding can occur if an insufficient number of point sources is used.

4.5.1 Z-Difference Penumbra Approximation

There are a number of fast shadow depth map algorithms [254, 66, 309, 597, 173, 338] that simulate the look of visually pleasing but approximate soft shadows from a point source. They lower the penumbra intensity (i.e., make the penumbra less dark) as a function of the distance from the occluder, but otherwise assume only a single point source. This makes sense because the farther away the occluder is from the point to be shaded, the more likely the point to be shaded is partially lit. However, this class of shadow algorithms will be noticeably incorrect for large extended lights.

The simplest approach [254] is to use a single Z-depth test: if in shadow, the difference between $\|P-L\|$ and Z_n is used as an indicator of the filter size to sample neighboring pixels (in the shadow depth map), and the softness of the penumbra is computed via percentage-closer filtering. This approach does not account for the outer penumbra and can produce light-leaking problems.

Fernando [173] introduces percentage-closer soft shadow depth maps (PCSS), where the penumbra size per point to be shaded is computed as

$$(\|P-L\| - Z_{\mathrm{avg}})\, W_{\mathrm{light}}/Z_{\mathrm{avg}},$$

where W_{light} is the width of the light and Z_{avg} is the average of the nearest Z-depths Z_n sampled (i.e., occluders). The penumbra size is then mapped to the filter size for which a PCF computation is done (see Figure 4.13 for rationale for the above equation and Figure 4.14 for a sample rendering). Note that the assumption of an averaged Z_{avg} may cause self-shadowing artifacts or improper penumbra sizes. Valient and deBoer [597] propose a similar technique.

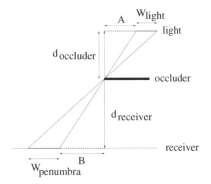

Figure 4.13. Computation of penumbra for percentage-closer soft shadows (PCSS).

More sophisticated schemes are available [66], in which for each point P to be shaded, if the Z-depth test shows it is in shadow, then P is considered as in shadow. However, if the Z-depth test shows it is illuminated, then the shortest distance to the closest shadow depth map pixel that indicates it is in shadow is identified. The penumbra intensity is a function of this shortest distance and the z-value difference between occluder and receiver. Note, however, this shortest distance can be limited to some maximum value so that the search does not become the bottleneck of the algorithm. Also, note that if only the inner penumbra were computed, the computation of the shortest distance is only needed for points in shadow. Another advantage of computing the inner penumbra is that a shadow-width map can be precomputed [309] to store the shortest distance information per shadow depth map pixel. Lawlor [338] extends the above approach for both

Figure 4.14. Shadow computations using percentage-closer soft shadow depth maps. *Image by Louis Bavoil, courtesy of NVIDIA Corporation.*

inner and outer penumbrae. The above approaches can be slower and require some knowledge or assumption of object relationships, which can get quite confusing with more complex scenes.

Among the above approaches, PCSS [173] is the cleanest and most often used approach. In fact, there are optimizations to this approach that combine PCSS with variance or convolution shadow maps (Section 2.3.6). They are appropriate extensions because of the potentially large filter sizes needed to achieve the soft shadows, which can significantly slow down the basic percentage-closer approach. Lauritzen [336] creates several prefiltered layers and stores them in a summed-area table format so that variable-size filters per point to be shaded can be generated. Dong and Yang [138] and Annen et al. [12] discuss a faster method for computing more accurate Z_{avg} for variance and convolution shadow maps, respectively.

Another avenue of performance improvement for PCSS is motivated by Robison and Shirley [486] via image-space gathering. In essence, PCSS information is stored in screen space, such that many fewer depth-map pixel samples are visited for the penumbra region (when projected to the screen), especially when the extended light is larger [417]. However, there is a loss in quality because the shadow depth map information is now discretized in screen space, and the filtering is trickier due to loss of edge information. The latter issue is improved by using a bilateral Gaussian filter [417] or an anisotropic Gaussian filter [672, 210]. It is unclear how much performance savings is gained using such techniques, especially with the extra preprocessing and memory needed, and it is unclear whether this might hinder integration with nonlinear shadow depth maps (Section 2.3.7).

4.5.2   Depth Values as Micropatches

Several independent papers [208, 29, 27, 40] treat each depth value as a square micropatch geometry (the size of the depth map pixel) parallel to the light. In determining a soft shadow value, the technique back-projects affected micropatches from the point to be shaded to the light source and queries all the shadow depth map pixels contained in the back-projection to get an accumulated area that represents the amount of shadowing (see Figure 4.15). Figure 4.16 shows a comparison between the micropatch and the penumbra wedges [5] approaches (Section 4.6.3) and flood fill [17] (Section 4.5.3).

Although this is a practical and fast approach, the generic approach has several limitations: the soft shadow computed does not entirely match the extended light, as the depth map information is due to a single shadow depth map from a spotlight. This is a generic limitation with this class of algorithms. However, there is progress to resolve other limitations, such as micropatch overlap artifacts (shadows are darker than they should be (see Figures 4.15 and 4.17)); micropatch gap (light-leaking problem) artifacts (Figure 4.15); performance can be slow due to the large number of shadow depth map values that need to be queried, especially for large extended lights.

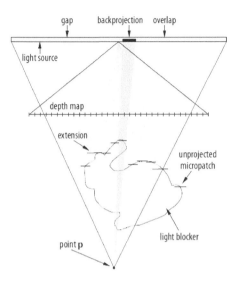

Figure 4.15. Using micropatches for shadow depth map soft shadows. *Image courtesy of Schwarz and Stamminger [511]. Computer Graphics Forum ©2007 The Eurographics Association and Blackwell Publishing Ltd. Published by Blackwell Publishing, 9600 Garsington Road, Oxford OX4 2DQ, UK and 350 Main Street, Malden, MA 02148, USA. Reproduced by permission of the Eurographics Association and Blackwell Publishing Ltd.*

To address the overlap issues, Schwarz and Stamminger [511] employ a jittered $n \times n$ pattern (by ray tracing) to get shadow occlusion, and Bavoil et al. [41] sample the back-projection region using a Gaussian Poisson distribution.

To address the gap issues, Guennebaud et al. [208] check the neighboring depth map information and extend the micropatches appropriately to the neighbors' borders. However, Bavoil [43] reminds us that bad self-shadowing artifacts will persist using micropatches. He uses mid-distance depth values and depth peeling to

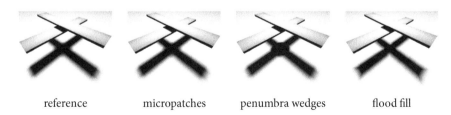

reference   micropatches   penumbra wedges   flood fill

Figure 4.16. Shadow computation comparisons. *Image courtesy of Guennebaud et al. [208], ©Eurographics Association 2006. Reproduced by permission of the Eurographics Association.*

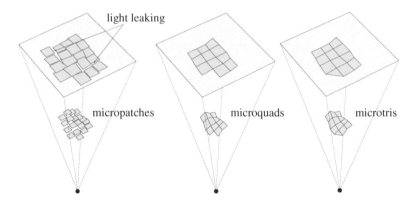

Figure 4.17. Back-projection with micropatches, microquads, and microtris.

reduce such artifacts, incurring additional performance cost. Alternatively, Schwarz and Stamminger [511] propose microquads, where each depth map pixel represents a vertex in the microquad. As a result, there are no gap problems, although performance is lowered due to the nonparallel back-projections. Extensions to the microquads include the microtris [512], and variable-sized microrects [513], to allow a better shape representation of the occluder. For the same reason, Guennebaud et al. [209] detect occluder contours from the micropatches, and perform radial area integration based on the occluded area; however, this approach may incorrectly miss occluded regions. Yang et al. [651] propose package-based optimization to speed up occluder contour construction. See Figure 4.17 for some of the different geometry types to represent micropatches.

To address the performance issues when a large number of shadow depth map pixels process large extended lights, Guennebaud et al. [208] create a mipmap of the shadow depth map, storing the *min* and *max* depth values. Any depth value less than the *min* value indicates full lighting represented by that shadow depth map region; any depth value larger than the *max* value indicates full shadowing represented by that shadow depth map region. However, the region represented by the mipmap is usually much larger than the actual extended light, which means this representation is not optimal. Schwarz and Stamminger [511] propose the multiscale shadow map, where each shadow map pixel at level i stores *min* and *max* depth values in a neighborhood region of size $2^i \times 2^i$ centered around that pixel. Because there is a better match of the accessed region versus the extended light, full lighting and full shadowing cases can be determined with less (level) accesses, but only by sacrificing extra memory usage. Schwarz and Stamminger [512] propose the hybrid Y shadow map to take the best of the above two approaches.

Some of the papers above discuss the concept of multilayered depth images [41, 511, 43, 425], of which the results are more accurate but at the cost of slower performance. Also, ray tracing variations of such an approach exist and will

light radius = 2   light radius = 4   light radius = 6

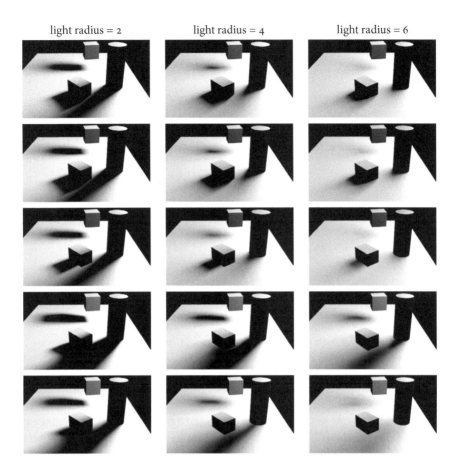

Figure 4.18. From top to bottom: distribution ray tracing [108], ray tracing multi-layer depth map [649], ray tracing single-layer depth map, multi-layered micropatches [41], and single-layered micropatches [208]. *Image courtesy of Xie et al. [649], ©Eurographics Association 2007. Reproduced by permission of the Eurographics Association.*

be discussed in Section 4.7.4. Light-leaking or gap problems exist for the ray tracing variations as well. In general, the ray tracing variations tend to provide more accurate soft shadows but are much slower than the algorithms discussed here (see Figure 4.18 for comparisons of various single and multilayered approaches).

4.5.3 Silhouette Detection-Based Solutions

Heidrich et al. [237] simulate soft shadows for linear lights. Two shadow depth maps, S_1 and S_2 are constructed at the endpoints of the linear light. If the depth

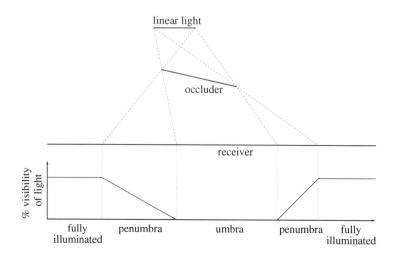

Figure 4.19. Visibility function of a linear light in the presence of an occluder.

comparisons for S_1 and S_2 indicate shadow, then the final illumination is 0 (fully shadowed). Equivalently, if the depth comparisons for S_1 and S_2 indicate no shadow, then the final illumination is 1 (fully lit). If depth comparisons indicate both shadowing and lighting, then L_i = visibility × local illumination, and the final illumination value for the linear light is $\sum_n L_i$, for n sample points on the linear light. The visibility interpolation (see the visibility function in Figure 4.19) is achieved through an associated visibility map by warping all the triangles from one view to another (and vice versa), and the shadow depth discontinuities are identified: this is known as the "skin," which produces the penumbra region (see

Figure 4.20. Resulting shadows for a linear light source (center), from skins generated for shadow depth maps. Skins generated for the left and right shadow depth maps are shown on both sides of the image. *Image courtesy of Heidrich et al. [237], ©Eurographics Association 2000. Reproduced by permission of the Eurographics Association.*

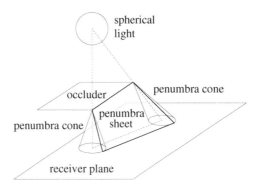

Figure 4.21. Penumbra cones and sheets are generated from the vertices and edge of the occluder.

Figure 4.20, where the skin for the shadows from the boxes are illustrated in the two shadow depth maps). When the light source is large, accuracy of the above approach may be poor, and the number of shadow depth maps generated can be increased above two so that the accuracy can be improved. Ying et al. [652] extend that approach for (polygonal) area lights in which a visibility map is computed per light boundary edge.

Wyman and Hansen [646] generate cones at silhouette vertices (Figure 4.21) and intermediate sheets along silhouette edges similarly to Haines [220] (Section 4.4.2). These volumes are then scan-converted in a penumbra map (see Figure 4.22(right)). For each cone or sheet pixel, the corresponding illuminated 3D point in the shadow depth map is identified. The shadow occlusion is $(Z - Z')/(Z - Z_n)$, where Z is the distance from the light to the point to be shaded,

Figure 4.22. A shadow depth map from the representative point source (left). The penumbra cones and sheets are scan-converted in a penumbra map (right). *Image courtesy of Wyman and Hansen [646], ©Eurographics Association 2003. Reproduced by permission of the Eurographics Association.*

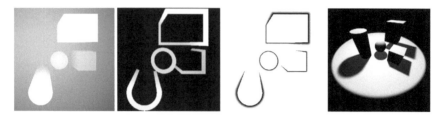

Figure 4.23. From left to right: shadow depth map from the representative point source, smoothie depth, smoothie alpha, and outer penumbra soft shadows. *Image courtesy of Chan and Durand [83], ©Eurographics Association 2003. Reproduced by permission of the Eurographics Association.*

Z_n is the distance from the light to the occluding vertex (or point on the silhouette edge), and Z' is the distance from the penumbra map. This method is suitable for hardware implementation and for interactive to real-time performance. Arvo and Westerholm [16] extend this approach to handle both the inner and outer penumbra regions.

Similarly, Chan and Durand [83] extend each silhouette edge of polyhedral models along two normals of the edge. This quadrilateral, called a "smoothie," is textured with an alpha ramp, simulating the penumbra produced by a spherical light source. These smoothies are then rendered (alpha and depth) from the light source in an alternate buffer (see Figure 4.23). If two smoothies fall on the same shadow depth map pixel, only its minimal alpha value (i.e., darker penumbra) is kept. In the final rendering, if a 3D point projects with a larger depth value than in the shadow depth map, it is considered in full occlusion. If it projects with a larger depth value than in the alternate buffer, its alpha value is used as an attenuation of the illumination. Otherwise, the 3D point is considered fully illuminated. Even though the shadows are only approximate and the algorithm only computes the outer penumbrae, the Chan and Durand technique still produces extended penumbrae of good visual quality at interactive to real-time rates. It also inherits many fundamental limitations of shadow depth maps, but extends their possibilities to produce antialiased soft shadows.

Along the lines of smoothies [83] and penumbra maps [646], Boer [61] introduces "skirts," which can produce both inner and outer penumbrae. The algorithm uses image processing techniques to detect silhouettes and then creates around a silhouette a skirt that has a width twice the radius of the light source, rendered into a skirt buffer. Another shadow depth map is generated with second-depth occluders, and the combination of the second-depth map and the skirt buffer generates soft shadows. Cai et al. [79] do something quite similar in what they refer to as "shadow fins."

Arvo et al. [17] use a single point source to simulate a spherical light and apply a flood-fill algorithm to achieve soft shadows. The scene with umbra shadows is first rendered using the standard shadow depth mapping techniques. A shadow silhouette algorithm is used to determine the silhouette of the umbra in the image. The penumbra rendering pass spreads out this umbra, where each pass spreads out by two pixels. The rendering passes terminate once the spreading is completed when the silhouette reaches the edge of penumbra. With these potentially many penumbra rendering passes, this algorithm appears to be quite slow, especially for larger light sources (additional optimizations using flood jumping are discussed by Rong and Tan [489]). However, overlapping objects will cause noticeable artifacts, as seen in Figure 4.16.

4.5.4   Multipoint Source Shadow Depth Map Algorithms

In the previous three sections discussing shadow depth map algorithms, these algorithms employ one or very few point light samples, which makes them very attractive in terms of real-time performance. In this section, many point light samples are relied upon to achieve more accurate results, but the algorithms tend to be slower. Although there is commonality in the approaches with respect to relying on multiple point sources, most approaches are really quite disjoint, i.e., the approaches are not based on one another.

Sung [570] identifies the triangles that reside in the volume defined by the point to be shaded and the rectangle encompassing the extended light. Those triangles are inserted into a hardware Z-buffer to automatically do the scan-conversion and determine the amount of occlusion, with the point to be shaded as the viewpoint, and the view direction as the center of the light. The scan-conversion is meant to simulate a set of faster distribution ray tracing computations. Note that this approach can be seen as an inverse shadow depth map approach, where the Z-buffer visibility is done from the perspective of the points to be shaded, not from the light-source origin.

Agrawala et al. [2] render shadow depth maps from different sample points on an area light source. These shadow depth maps are warped to the center of the light source and combined into a layered depth image, where each pixel in the layered depth image consists of depth information and a layered attenuation map, which stores the light attenuation at each depth. During display, each visible point to be shaded is projected onto the layered depth image and the attenuation is computed by comparing the depth value with each layer. Note that ray tracing variants of the layered shadow depth map are described in Section 4.7.4, but they are much slower.

St-Amour et al. [555] also warp each sampled point light to the center of the extended light as in the above approach. However, the information is stored within a deep shadow map [362] (Section 5.4.2 for a detailed description) instead of a layered depth image. This approach, called penumbra deep shadow mapping, also allows the combination of semitransparency, soft shadows, and motion blurred shadows, all dealt with in a single algorithm. However, the approach results in a very large structure where the extended light and the occluders are assumed static. Soft shadows can be cast over moving objects, but the objects themselves cannot

Figure 4.24. Extensions of the deep shadow depth map that include soft shadows. *Image courtesy of St-Amour et al. [555].*

cast soft shadows computed from this structure. See Figure 4.24 for a sample rendering.

Schwarzler [514] uses an adaptive sampling approach to create shadow depth maps for each frame. Four shadow depth maps are initially created, each point representing a point on the corner of the rectangular light. The shadow depth maps are compared to determine if further inner samples are necessary, and if so, further shadow depth maps are generated and so on. The criteria for determining the need for additional shadow depth maps can be tricky, and if not determined correctly, shifts or jumps in an animated sequence will surely appear.

Additional multipoint light-source shadow depth map techniques can be found in the following research [267, 268]. The two most interesting techniques in this section remain the layered attenuation map [2] and the penumbra deep shadow map [555].

4.5.5   Rendering Large Scenes

None of the shadow depth map algorithms that generate soft shadows that we have discussed thus far address the need to deal with artifacts due to the lack of focused resolution. In other words, it is not uncommon for the above approaches to exhibit sawtooth-like results in large scenes. Only a couple of approaches have been discussed in the literature for large scenes, such as extensions to alias-free shadow maps and fitted virtual shadow depth maps. We do find it curious why the combination of PCSS (Section 4.5.1) with z-partitioning (Section 2.3.7) have not been proposed to date, as this seems the most obvious extension.

Alias-Free Shadow Maps

The alias-free shadow map (or irregular Z-buffer) algorithms [535, 276] described in this section are soft shadow extensions of their hard shadow equivalents [4, 275] described in Section 2.3.7. In general, for the visible pixels, they first calculate the umbra, then use various techniques to approximate the penumbra.

Sintorn et al. [535] first compute the camera view pass to determine which pixels need to answer the shadowing question. Each triangle is then processed, in a single pass similar to the one from Laine and Aila [325], to compute the conservative influence region of the umbra and penumbra from each triangle (see Figure 4.25 for construction of the conservative influence region). Instead of a single bit as attached to the alias-free shadow map in the hard shadow case, a bitmask is associated with each visible pixel to indicate visibility of (typically) 128 light samples. Temporal jittering of each sample can also be done to account for motion blur.

Johnson et al. [276] also compute the camera view pass to determine which pixels need to answer the shadowing question. Those visible pixels are stored in a 3D perspective grid (see Figure 4.26). Instead of a bitmask approach to compute soft shadows, they perform geometric projection of the triangles to capture umbra information and create penumbra wedges (as in [23]; see Section 4.6.3) for silhouettes only relevant to those visible pixels. This is especially significant for the penumbra-wedge approach since its main drawback is complexity, which is dramatically reduced in this technique. Additional optimizations for this approach are discussed by Hunt and Johnson [253].

The approach of Johnson et al. [276] has the advantage of a geometric approach (using penumbra wedges) but with reduced cost. It generates soft shadow solutions analytically. It has inherent disadvantages of penumbra wedges such as incorrect overlap results. The approach of Sintorn et al. [535] does not have the incorrect

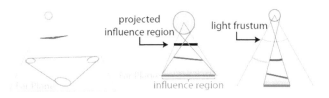

Figure 4.25. Computing the conservative influence region for the alias-free shadow map. *Image courtesy of Sintorn et al. [535]. Computer Graphics Forum ©2008 The Eurographics Association and Blackwell Publishing Ltd. Published by Blackwell Publishing, 9600 Garsington Road, Oxford OX4 2DQ, UK and 350 Main Street, Malden, MA 02148, USA. Reproduced by permission of the Eurographics Association and Blackwell Publishing Ltd.*

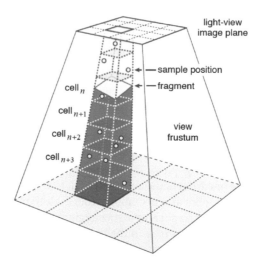

Figure 4.26. 3D perspective grid to store visible pixels. ©*2009 ACM, Inc. Included here by permission [276].*

overlap situation, but may likely need more computation time due to the need for 128 samples per visible pixel—this can turn into an advantage with temporal jittering to account for motion blur.

Both approaches currently do not provide real-time performance in general, but remain an interesting approach for further research.

Fitted Virtual Shadow Depth Maps

Shen et al. [522] use a variation of fitted virtual shadow depth maps (Section 2.3.7) and percentage-closer soft shadows (PCSS—see Section 4.5.1). First a low-resolution shadow depth map is rendered to allow the estimation of the penumbra width of each visible point. This information is then used to predict the required resolution of a depth map tile in a fashion similar to fitted virtual shadow depth maps. High resolution depth map tiles are rendered, and PCSS is applied for the final rendering.

This approach is very reliant on the accuracy of the tile resolutions, or else visible artifacts between the tile borders (due to different resolutions) can be visible.

4.5.6   Other Shadow Depth Map Variations

Mo et al. [416] introduce the concept of a depth discontinuity occlusion camera (DDOC), which is a non-pinhole camera that is placed in the center of the extended light. A distortion shadow depth map is created, where the distortion occurs around the discontinuities or silhouettes of the objects. The distortion causes bending of the shadow rays and thus generate soft shadows. While this approach is theoretically interesting, the creation of the distortion map is very slow.

Jakobsen et al. [263] store multiple z-values per pixel to indicate the transition of the umbra to penumbra regions. However, this algorithm suffers from projection aliasing problems, where the slightest bias or offset values can cause incorrect self-shadowing.

Ritschel et al. [479] introduce coherent shadow depth maps in which a set of many shadow depth maps per object are precomputed and compressed, each set containing per-object visibility. Thus, with any animation where the objects' convex hulls do not overlap, a distributed sampling approach to compute soft shadows can be very fast. However, this approach implies significant preprocessing time and memory overhead.

Scherzer et al. [506] generate a single shadow depth map based on stochastically choosing a point on the extended light. Shadow determination is based on previous frames' shadow depth maps, which represent other points on the extended light and the current frame's shadow depth map. Some smoothing is necessitated by the flickering that results from this approach, and likely additional self-shadowing artifacts will appear.

4.6   Shadow Volumes

There are three basic shadow volume approaches—the multipoint shadow volumes (see Section 4.6.1), shadow volume BSP (SVBSP; see Section 4.6.2), and penumbra wedges (Section 4.6.3). The penumbra-wedge algorithms have received significantly more research attention even though the penumbra wedge approach cannot easily use many of the culling techniques of shadow polygons as discussed in Section 2.4.3, especially when a hard boundary for a silhouette from an extended light does not exist. In comparison, the SVBSP approach has not been attempted on the GPU and has the tree structures growing very quickly in size and complexity. The multipoint shadow volumes approach is the most accurate, because it attempts to simulate point sources of the extended light, but is also the slowest approach.

4.6.1   Multipoint Shadow Volumes

Brotman and Badler [71] stochastically choose points to model extended light sources. Their algorithm generates shadow polygons (in the form of shadow volumes) for each point source. A 2D depth buffer for visible surface determination is extended to store cell (pixel) counters. The cell that the point to be shaded resides in is found, and the associated cell counter is incremented by 1 if the shadow polygons for that point source enclose the whole cell. If the corresponding cell count equals the total number of point sources, then the point to be shaded is in full shadow. If the cell count is less than the total number of point sources but higher than 0, then the point lies in penumbra. Diefenbach and Badler [128] and

Udeshi [595] implement the above approach on the GPU using multipass OpenGL rendering.

4.6.2   Shadow Volume BSP (SVBSP)

Chin and Feiner [92] extend their own work on SVBSP data structure [91] (Section 2.4.4) to achieve soft shadows. Instead of a single BSP shadow tree, as in their original paper [91], two shadow trees are created: one umbra tree and one penumbra tree, which encompasses both penumbra and umbra. The polygon to be shaded first explores the penumbra tree. If it reaches an *out* cell in the penumbra tree, then the polygon is fully lit. If this same polygon reaches an *in* cell, then it may be in umbra or in penumbra. This ambiguity is resolved by exploring the umbra tree. If it reaches an *out* cell, then the polygon is in penumbra; otherwise it is in full shadow.

The intensity of the penumbra for the SVBSP tree is based on contour integration for diffuse receiving surface elements. The light source may also need to be partitioned so that a unique BSP-tree traversal order will be generated, i.e., so that the split up area light will be entirely on one side of partitioned planes. Wong and Tsang [636] indicate that this can be very wasteful and identify cases where this extra partitioning of the area light can be avoided.

4.6.3   Penumbra Wedges

Akenine-Möller and Assarsson [5] modify the definition of a shadow volume to produce soft shadows. Instead of a quadrilateral forming a shadow polygon, a *wedge* is formed from a silhouette edge that represents the penumbra region. Each wedge has a shape similar to a prism, and all the wedges surround the umbra (see Figure 4.27). To determine visibility, instead of incrementing and decrementing a

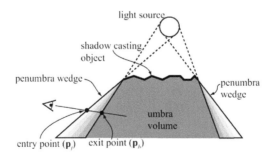

Figure 4.27. Penumbra wedges with penumbra intensity variations within a penumbra wedge. *Image courtesy of Akenine-Möller and Assarsson [5], ©Eurographics Association 2002. Reproduced by permission of the Eurographics Association.*

shadow count, a fractional counter (of changing lighting intensity) is used to increment and decrement when entering and exiting the wedges. A higher-precision stencil buffer is needed to deal with this lighting intensity counter. The intensity of the penumbra region has a linear decay between the wedge boundaries. As a consequence, the penumbra intensities are approximate, and when wedges of a single silhouette edge cross or are contained in one another, the shadow results can be wrong.

Assarsson and Akenine-Möller [24] improve on the original algorithm's performance by separating the umbra and penumbra rasterizations. The umbra computations are virtually identical to the typical shadow volume approach, using the center of the light as the umbra point light source. The penumbra is only of interest if the point to be shaded resides within a wedge (i.e., in penumbra). If within a wedge but not in shadow from the umbra pass, then the shadow fraction is decreased; if within a wedge but already in shadow from the umbra pass, the shadow fraction is increased to reduce the hardness of the umbra. Separating the two rasterizations allows less intersection computations than in their original work [5], and since only visible surface points that reside within a wedge are of interest, the bounding box of minimum and maximum Z-depths can be used to cull a lot of penumbra rasterizations.

Despite the speed improvements [24], the above approaches can cause artifacts when the wedges of a single silhouette edge cross or are contained in one another; thus, another variant of the penumbra wedge solution has been proposed [23] where wedges act as conservative penumbra regions that are combined in a visibility buffer bitmask to arrive at the shadow solution. In fact, the wedge created is independent of the light silhouette, and only the center point of an extended light is considered. The penumbra produced is fairly accurate unless the size of the light source is large in comparison to its shadow casters, which then results in overstated umbrae. Forest et al. [175] improve on the above problem by splitting the light into four regions, optimized by doing it in a single pass.

In terms of improving performance for the above, Assarsson et al. [25] include optimizations such as tighter shadow wedges and an optimized GPU pixel shader for rectangular and spherical lights. Viney [604] constructs penumbra wedge geometry directly on the GPU using geometry shaders. Lengyel [346] suggests treating a penumbra wedge as two halves, containing an inner and outer penumbra. An optimization would be to only process the outer penumbra wedge, resulting in less accurate results that still look good.

Forest et al. [176] use penumbra wedges and depth complexity sampling to arrive at a real-time soft shadow solution. See Section 4.7.3 for a ray tracing implementation of the general approach. Forest et al. apply that general approach but use a sampling grid of only 16 samples, and instead of using ray tracing to compute the reference sample, they apply a shadow volume approach. See Figure 4.28 and note the overstated umbrae from the penumbra-wedge approach, as compared to the more accurate result from Forest et al. [176]. Although penumbrae are generally

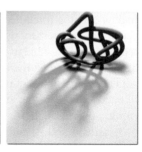

Figure 4.28. Comparing soft shadows from the following: penumbra wedges (left); the approach of Forest et al. (center); reference image (right). *Image courtesy of Forest et al. [176]. Computer Graphics Forum ©2008 The Eurographics Association and Blackwell Publishing Ltd. Published by Blackwell Publishing, 9600 Garsington Road, Oxford OX4 2DQ, UK and 350 Main Street, Malden, MA 02148, USA. Reproduced by permission of the Eurographics Association and Blackwell Publishing Ltd.*

much improved, the approximate precision of the shadow fraction of the shadow volume approach may still result in penumbra errors.

Note that the implementations for the above penumbra wedge algorithms tend to be quite complex due to the many cases that need to be properly handled, and unfortunately, very little of the code from shadow volume hard shadows can be reused for soft shadows. The reward for the above algorithms remains the visually pleasing rendered results.

4.7   Ray Tracing

Ray tracing remains the most accurate, flexible, but also the slowest approach to achieving soft shadows unless the platform supports a high level of parallelism. Most of the research resides in distribution ray tracing (Section 4.7.1), although some work has been done in analytic solutions through back-projections (Section 4.7.2). Distribution ray tracing approaches tend to be easier to implement, but getting high-quality results with minimal noise and without resorting to many point light source samples tends to be challenging. Analytic solutions are more difficult to implement, but they often provide smoother-looking results. The main decision criteria to choose between the basic approaches is fundamentally how many point samples can match the analytic solution in terms of performance and quality. This is a difficult question to answer, although it is appropriate to conclude that the larger the size of the extended light, the more feasible analytic solutions become.

Faster alternatives to the above include the single-ray analytic solution (Section 4.7.3) and ray tracing depth images (Section 4.7.4), although care must be taken to avoid artifacts when using ray tracing depth images.

4.7.1   Distribution Ray Tracing

In terms of soft shadows within ray tracing, variations of the distribution ray tracing approach [108] have become the most often used technique. An example
is the work using distribution ray tracing for linear and area light sources [463]
and curve light sources [33]. By distribution ray tracing, we mean shooting a
number of rays towards the extended light to come up with an average for the
shadow occlusion fraction or to estimate the radiance at each sample. This does
not imply that it always requires stochastic sampling as implemented in the original work of Cook et al. [108]. Deterministic sampling approaches (e.g., adaptive supersampling [629], stratified supersampling [340], Hammersley and Halton point sequences [637], N-rooks with a static multijittered table [617], and
blue-noise tiling patterns [442], etc.) for distribution ray tracing can work just as
effectively.

An important aspect in favor of distribution ray tracing [108] is that noise is
far more visually acceptable than banding, and when viewed over a sequence of
frames, the noise is hardly noticeable. Thus, stochastically shooting shadow rays,
hitting different points on the extended light over a sequence of frames, can achieve
good results. Other variations of the distribution ray tracing techniques are listed
below, mainly focusing on how to reduce the number of shadow rays while maintaining good quality results.

For the rest of this section, a set of disjoint papers are presented. They have
little in common except that they discuss soft shadows computed by a distribution
of rays.

Shirley et al. [527, 621, 528] compute good estimators of probabilistic locations
on the extended light source for lighting and shadow computation used within a
distribution ray tracing environment. Ideal probabilistic locations are discussed
for various types of extended light sources. See a more detailed description of this
work in Section 2.5.3.

Jensen and Christensen [269] apply a preprocessing step for computing a photon map [273] by sending photons from the light source. For each photon intersecting a surface, a shadow photon is continued along the same direction as if the
original photon had not hit any surface. To compute the illumination at a particular
point P, the nearest photons around P are gathered. If they are all shadow photons,
then P is considered in complete shadow. If they are all regular photons, then P is
fully lit. If they consist of both regular and shadow photons, then P is either on the
boundary of the shadow or it is in the penumbra of an extended light. One can take
the number of regular and shadow photons to determine the shadowing fraction,
but the authors [273] indicate that the result is not very good unless a very large
number of photons are generated from the light. Instead, shadow rays are shot to
the light source to determine shadowing. In essence, the photon map is applied
as a shadow "feeler." Using this technique, the authors claim that as many as 90%
of shadow rays do not need to be cast. However, the preprocessing cost may be

large for scenes only directly illuminated, and because the photons are shot based on probabilities, small objects can be missed.

Genetti et al. [189] use pyramidal tracing to quickly compute soft shadows from extended lights. A pyramid from the point to be shaded to the extended light is formed. If there are different objects in this pyramid, then the pyramid is subdivided into smaller pyramids as in typical adaptive supersampling approaches [629]. The subdivision criteria remain the same geometric identification the authors use in their earlier work [188], which means that the same limitations still hold, i.e., the criterion of the presence of different polygons is used to determine how complex that pyramid region might be and thus drives how many rays need to be shot. This may be ideal for higher-order surfaces, or large polygons, but not ideal for polygonal meshes, as they will always require many shadow rays to be shot based on the above criterion.

Hart et al. [225] precompute an occluder list per image pixel per light source by tracing a very small number of shadow rays. When an occluder is found, a check is done to determine if adjacent image pixels also "see" the same occluder so that the adjacent pixels can also register this occluder. During the shading of the pixel, the pixel's occluders are projected and clipped against the extended light to analytically determine the visible portions of the light. This algorithm ensures consistency of occluders between adjacent pixels but can result in missed occluders (i.e., light-leaking problems), especially for small geometry occluders. The storage of occluders per pixel can also be large.

Distribution ray tracing has enjoyed much faster performance with SIMD instructions, parallel systems on either the CPU or GPU, and ray packets for locality of reference [64, 45]. Another optimization exploiting locality of reference can be applied to the geometry itself, using hierarchical penumbra casting [325]. In this approach, all shadow rays are identified first, then a loop through each triangle occurs where the algorithm finds all shadow rays that intersect this triangle.

Additional reading on other distribution ray tracing optimizations includes

- Extensions of the hemicube data structure towards area sources [407].

- Extensions of the mailbox/rayID optimization to reduce computation of the non-first shadow ray [643].

- Approximate contour integration to compute illumination from diffuse area sources using distribution ray tracing only for the occluded or partially occluded cases [603].

- Use of the Minkowski operators [178, 372] and frequency analysis [148] and gradients (normals) analysis [473] to focus the regions where shadow rays need to be shot to get the desired details. Additional advanced frequency analysis has been researched [154, 153].

- Prefiltering and storing local occlusion information. Shadow rays can terminate earlier based on ray differential and the local occlusion information [321].

4.7.2 Structures for Exact Back-Projection

Amanatides [10] extends the ray tracing concept of a ray to a cone. Instead of point sampling, cone tracing does area sampling. Achieving antialiasing requires shooting exactly one conic ray per pixel. Broadening the cone to the size of a circular light source for shadow cones permits generation of soft shadows: a partial intersection with an object not covering the entire cone indicates penumbra, an intersection with an object covering an entire cone indicates umbra. Due to locally complex scenes, it may be necessary to divide up a single cone into a set of smaller cones to get a better approximation for soft shadowing, which can be complicated. Note that it is ideal to use cone tracing only for spherical (or ellipsoidal) lights. The lack of adoption of cone tracing is likely due to little success in performing efficient intersection culling. Figure 4.29 illustrates how cone tracing generates soft shadows.

Heckbert and Hanrahan [231] introduce beam tracing, which is very similar to cone tracing, except that beams (elongated pyramids) replace cones. It has the advantages of cone tracing, such as achieving soft shadows naturally (in this case, for polygonal lights). However, it also has the disadvantages of cone tracing, such as complex beam-geometry intersections and lack of an effective culling structure. Overbeck et al. [447] introduce fast algorithms for ray-geometry intersection and k-d tree traversal as a culling structure, which has pushed the performance of beam tracing quite significantly.

Poulin and Amanatides [465] introduce two structures to soft shadows for a linear light. The first involves a light triangle, where a light triangle is defined by the endpoints of the linear light and the point to be shaded. The regular grid

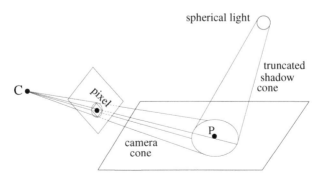

Figure 4.29. Approximating the shadowing of the pixel-cone intersection by a spherical light shadow cone.

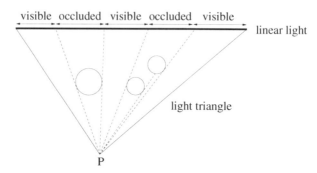

Figure 4.30. The light triangle formed by the point to be shaded and the linear light is intersected against the objects of the scene. The intersected objects are back-projected onto the linear light to determine the light illuminating segments.

voxels that encompass this triangle are identified through 3D scan-conversion, and the objects in these voxels are intersected against the light triangle. All objects that intersect the light triangle are projected toward the linear light to determine occlusion. Although this algorithm provides an analytic solution to the shadow value, the authors indicate that the results are slow due to the expensive scan-conversion. In fact, this analytic solution is one of the few that generate the correct soft shadow integral evaluation. See Figure 4.30, which shows the light triangle and how the shaded regions are the final segments of the linear light that are fully lit, and see Figure 4.31 for a sample rendering.

Figure 4.31. Analytic solution for soft shadows from two linear light sources per rectangle. *Image courtesy of Poulin and Amanatides [465], ©Eurographics Association 1990. Reproduced by permission of the Eurographics Association.*

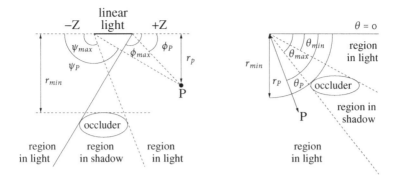

Figure 4.32. Angles ϕ and ψ from the linear light axis [577].

Poulin and Amanatides [465] also propose a more efficient linear light buffer represented by an infinite cylinder oriented along the light. The cylinder is subdivided into arcs along its radius. Objects radially projecting in an arc are added to the arc's list. Any point to be shaded identifies the arc it resides in and projects the list of objects associated with this arc on the linear light. To further reduce arcs, each arc is divided into three parts: the left, center, and right side of the linear light. They [465] found this algorithm to be much faster than the scan-conversion approach, although it does require additional preprocessing and more memory to store pointers to the entire object set.

Tanaka and Takahashi [577] extend the linear light buffer so that the list of objects is partitioned as in the previous scheme [465], but it is also partitioned into layers along the radius of the infinite cylinder as well as in overlapping, partitioned bands that are ϕ and ψ angles from the linear light axis (see Figure 4.32). The bands represent regions for potential shadow rays. A loop through all the partitioned regions is executed, projecting all hit segments with the light triangle to get the final shadow occlusion value. The memory requirements for this scheme appear to be quite large.

The same authors [578] extend their approach to deal with area lights. Partitioned bands are created in each direction of the area light, and a 2D array of bands is produced. The light pyramid (instead of light triangles for linear lights) is intersected against the objects in the appropriate band. Instead of employing a typical and expensive polygon clipper to identify the lit regions on the area light in order to compute the visible regions of the area light analytically, a cross-scanline clipping algorithm [576] is used. Both algorithms provide analytic solutions to the shadowing of extended lights. Similarly, Stewart [563] is also able to compute exact regions of visibility with the area light source and the point to be shaded. Because of rendering systems that supersample to antialias, it is unclear if distribution ray tracing [108] with a few shadow rays per supersample converges to the analytic solution just as easily without incurring the cost of the exact analytic solution.

Another class of back-projection techniques have been applied to radiosity [560, 141, 561, 142, 368, 101, 32] in order to find an optimal discontinuity mesh (see Section 5.10.1 for details).

4.7.3   Single-Ray Analytic Solution

Parker et al. [452] apply ray tracing to generate approximate but pleasant-looking soft shadows by shooting only one shadow ray per spherical light. The basic idea is to identify the silhouette of objects and make semitransparent a region around the silhouette. These semitransparent regions therefore produce penumbrae. Any penumbra region outside the umbra results in some interpolation from the umbra to achieve soft shadows, i.e., computation of the outer penumbra. Note that the umbra tends to be overstated, and a spherical light is assumed. This approach was also the basic motivation for non-ray tracing approaches for soft shadow approximation [220, 66, 83, 309]. See Figure 4.33 on the approach and Figure 4.34 for a sample rendering.

Laine et al. [328] use penumbra wedges (Section 4.6.3) and depth complexity sampling to accelerate the computation of accurate shadows by shooting a single shadow ray per extended light. This is achieved by extracting silhouette edge information and storing it in an acceleration structure such as a hemicube. With a shadow visibility query, a list of penumbra wedges of all potential silhouette edges is created by projecting the point to be shaded onto the surface of the hemicube. There are a fixed number of light samples (up to 1024 samples) representing an extended light, where each such light sample will accumulate the depth complexity. The depth complexity of the light sample is incremented if an edge is projected onto that light sample. A single shadow ray is then cast toward the light with the lowest

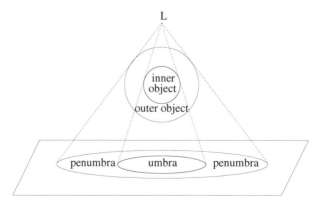

Figure 4.33. A region around the object is marked as semitransparent. Any ray traversing this region is considered partly occluded, depending on its distance to the real object.

Figure 4.34. Hard shadows from a single point (left). Soft shadows approximated from inner and outer object regions (right). *Image courtesy of Parker et al. [452].*

depth complexity. If the shadow ray is blocked, then the point is in full occlusion; otherwise, the light samples with the same lowest depth complexity are fully visible. Further work on the above general approach has been focused on an acceleration data structure to replace the hemicube, such as k-d trees [226] and 3D grids [343]. The basic approach of depth complexity sampling should be much faster than distribution ray tracing except that it is unclear, when the geometric complexity (i.e., number of edges) of the scene is large, whether the $O(n)$ geometric behavior of the above algorithm will provide performance gains.

4.7.4   Ray Tracing Depth Images

Lischinski and Rappoport [352] achieve computationally faster soft shadows by hierarchically ray tracing layered depth images (LDI) instead of the slower ray tracing of scene objects. A shadow LDI is generated that represents the view from the center of the extended light source. By itself, LDIs can result in missed shadow hits, very similar to the early implementation of micropatches (Section 4.5.2). To alleviate this, each depth is superimposed onto fixed boundaries, thus forming discrete depth buckets [297], resulting in larger, flat surfaces (e.g., panels). To improve performance, a 32-bit word can be used to describe 32 depth buckets at a pixel. Shadow rays traverse the depth buckets (to the extended light), and the bitmasks are ORed together to determine shadow visibility for each shadow ray (see Figure 4.35). For shadow depths that may cause self-shadowing, the original Z-depth values are kept around for comparison. Although this scheme is fast and reduces the risks of missed shadow hits, light-leaking problems can still occur, and the discrete depth buckets may show shifting artifacts over an animation. Also, the soft

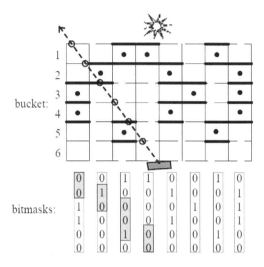

Figure 4.35. The scene is converted in a layered depth image. The soft shadows result from tracing shadow rays through this structure and testing for intersection with the flat panels in occupied voxels. *Image courtesy of Keating and Max [297], ©Eurographics Association 1999. Reproduced by permission of the Eurographics Association.*

shadows are very approximate because the LDI represents a single view from the center of the extended light, and shadow rays are shot to the extended light. As a result, the size of the extended light tends to be small for the results to be visually acceptable. Im et al. [256] extend the above approach by using depth peeling to generate the LDIs. Xie et al. [649] accelerate the above approaches by storing a quadtree of min_z and max_z as well as opacity values per layer to achieve semitransparent shadows as well. Visual comparisons of this approach versus micropatches (Section 4.5.2) on soft shadows can be seen in Figure 4.18. If Xie et al. [649] allow an unlimited number of layers, one can consider the approach as being a data structure suitable for volume graphics as well (Section 3.7).

Agrawala et al. [2] deal with the light-leaking problem in a different way, using an approach that is similar to ray tracing heightfields. Multilayer depth images are generated from points emanating from the extended light; thus, this approach is more accurate than the above approaches, which assume only a single point from the extended light. To compute shadowing for each depth image, an epipolar ray traverses the depth image. As the traversal occurs, the epipolar depth interval $[Z_{enter}, Z_{exit}]$ is maintained. If the image depth Z_{ref} is inside the epipolar depth interval, then it is in shadow. To speed up traversal, a quadtree is used. Because a fixed set of depth images are preprocessed, if the number of such depth images is too small, banding artifacts in the soft shadows may occur.

4.8   Other Soft Shadow Algorithms

Yoo et al. [653, 654] provide fast algorithms to compute the exact umbra and penumbra events of a single polygon, which is exploited by Bala et al. [32] to achieve fast antialiasing results without supersampling. Stark et al. [558] compute the exact shadow irradiance using polygonal splines. Demouth et al. [127, 126] compute the exact polygonal boundaries between umbra and penumbra sections. Boehl and Stuerzlinger [60] merge geometric silhouettes and filter the results from several point light sources (representing the extended light). As it currently stands, the derived mathematics for the above papers are very interesting but less practical in complex environments. The results could be input towards discontinuity meshing in radiosity (Section 5.10.1).

Ouellette and Fiume [445, 446] determine the approximate locations of integrand discontinuities for points in the penumbra. These points correspond to the approximate boundaries of the visible portions of the linear light. The authors take advantage of the fact that even with complex occlusions, there are usually only one or two such discontinuities. They demonstrate that knowing the exact locations of these points is not necessary, provided that the average location of these discontinuities varies continuously throughout the penumbra. They introduce the Random Seed Bisection algorithm, a numerical iterative technique for finding these discontinuities in a continuous fashion. The same authors [444] follow up with research to locate integrand discontinuities for area lights. The visible domains of the integrand are approximated by a polygonal boundary, then subdivided into triangles so each of the triangulated integrands can be computed analytically using a low-degree numerical cubature (third degree analogy to a numerical quadrature technique).

Donnelly and Demers [139] use occlusion interval maps to produce soft shadows on the GPU that are feasible only for static scenes and a fixed light trajectory over time (e.g., a sun). An occlusion interval map is very similar to a horizon map (Section 5.2) in which each point has a hemisphere of information in terms of occlusion, computed using ray tracing. The occlusion function is usually a bar chart of binary-decision visibility occlusion, and can be turned into a smoother function to achieve soft shadows. During real-time playback, the shadow information is simply retrieved from this map. There are hints of ambient occlusion in this approach (see Section 5.8).

4.9   Trends and Analysis

From a research paper standpoint, it is very obvious that shadow depth maps have captured most of the attention, ahead of shadow volume variations. This is likely due to the flexibility and performance capabilities of shadow depth maps. Ray tracing has also generated a lot of interest because it computes the most physically accurate shadows without fail and because of the speed improvements gained in the last decade. Most of the speed improvements have positively affected the single-ray analytic solution, distribution ray tracing, and beam tracing.

Although the number of papers on soft shadows from direct illumination has been large, they pale in comparison to the number of papers published for hard shadows. The current count for the papers is

Planar receivers of shadows	7
Shadow (depth) maps	51
Shadow volumes	9
Ray tracing	42

4.9.1   Offline Rendering Trends

In offline rendering situations for turnkey solutions, distribution ray tracing [108] is almost universally available to generate soft shadows from extended lights. Shadow depth maps are still the preferred solution for performance reasons, but there is very little attempt to offer correct soft shadows from shadow depth maps, as the PCF blurring [476] or its extended light version, PCSS [173], tend to be good enough in many cases. This combination is also powerful in that there is much common code to support hard and soft shadows (thus reducing code maintenance).

4.9.2   Real-Time Rendering Trends

In real-time turnkey engines, extended lights are seldom offered, and correct soft shadows are almost never available aside from PCF blurring [476]. From a shadow depth map perspective, we believe this is because no one has pieced together the ability to render large scenes in real time without exhibiting sawtooth-like results. Section 4.5.5 discusses those papers that come the closest to achieving this.

From a penumbra wedge perspective, performance remains a bottleneck, especially when many of the culling techniques of the shadow polygons as discussed in Section 2.4.3 cannot be easily applied (mainly due to a lack of a hard boundary silhouette). It also feels like a robust implementation is required to handle many special cases, with very little code in common with hard shadow volumes.

The lack of adoption of correct soft shadow generation from extended lights may also be due to focus on other sources of soft shadows, including indirect illumination (ambient occlusion (Section 5.8), precomputed radiance transfer (Section 5.9)), and global illumination (Section 5.10).

4.9.3   General Comments

When soft shadows from extended light sources are required, it is rare that only soft shadows are offered. Rather, there is usually an option to achieve hard and soft shadows. Thus, due to reduction of code maintenance and consistent user expectations, the tendency is to offer rendering solutions of the same type in both hard and soft shadows. For example, one may prefer to offer some shadow depth map variation in both hard and soft shadows instead of shadow volumes for soft shadows and shadow depth maps for hard shadows. We mention this because the choice of a soft shadow rendering solution could be dictated by both hard and soft shadow choices. The same can be said for considerations for semitransparent shadows, motion blur, and atmospheric shadows, which are discussed in the next chapter. This is why a technique such as deep shadow maps [362] has been used often, as it is able to offer hard, soft (Section 4.5.4), semitransparent (Section 5.4.2), complex thin material (Section 5.5), and motion-blur (Section 5.7) shadows.

A perspective of algorithmic choices can be seen in the following four categories, quantifying the pitfalls, quality, and performance, so that the reader can decide what is acceptable for his needs:

1. *Single point source.* This class of algorithms produce smooth soft shadows with a single point. They are generally the fastest but also the least physically accurate. A few of these algorithms exhibit light-leaking problems.

2. *Boundary/silhouette cases.* This class of algorithms produce smooth-looking shadows due to extraction of boundary or silhouette cases and are based on sampling one or very small number of point sources. However, they have artifacts from geometry overlap, and sometimes only the outer penumbra is considered. A few of these algorithms also exhibit light-leaking problems.

3. *Multiple point sources.* This class of algorithms represent the most physically correct soft shadows from extended lights. They can converge to smooth-looking shadows if enough point sources are used (performance might be slow); otherwise, banding or noise may occur. This is usually the slowest category of algorithms if a good visual result (minimal noise or banding) is desired, although these algorithms have no issues with geometry overlap nor light leaking.

4. *Analytic solution.* Alternative to multiple point sources, this technique arrives at a solution analytically. This class of algorithms produce smooth-looking shadows that are usually physically accurate. Performancewise, they are usually between the algorithms of categories (2) and (3).

Table 4.1 is motivated by a similar table from Johnson et al. [276], and as can be seen, the four categories proposed in this chapter appear to support useful trends, such as all boundary/silhouette-based solutions having issues around accurate geometry overlaps. Note that the lack of inner penumbra support is only an issue of accuracy, while accurate overlap and light leaking affect accuracy and consistency (i.e., easily visible artifacts), which are more significant. Also note that all

Chapter.Section + Algorithm	Inner penum.	Accurate overlap	No light leaking	Light type
(1) Single point source				
4.5.1 PCSS	✓	✓	✓	sphere
4.5.2 Micropatches	✓	maybe		rect/sph
4.5.2 Occluder contour	✓		✓	rect/sph
4.5.2 Microquads/microtris	✓	✓	✓	rect/sph
4.7.3 Ray trace single sample		✓	✓	sphere
4.7.4 Ray trace LDI	✓	✓	maybe	sphere
(2) Boundary/silhouette				
4.4.2 Plateau shadows			✓	sphere
4.5.3 Heidrich linear light	✓		✓	segment
4.5.3 Penumbra map	maybe		✓	sphere
4.5.3 Smoothies				sphere
4.5.3 Flood fill				sphere
4.5.5 Irregular Z-buffer	✓		✓	sphere
4.6.2 SVBSP	✓		✓	polygon
4.6.3 Penumbra wedges	✓		✓	rect/sph
(3) Multiple point sources				
4.4.1 Soft shadow textures	✓	✓	✓	polygon
4.5.4 Layered attenuation map	✓	✓	✓	sphere
4.5.4 Penumbra deep shadow map	✓	✓	✓	sphere
4.5.5 Alias-free shadow map	✓	✓	✓	sphere
4.7.1 Distributed ray tracing	✓	✓	✓	any
4.7.3 Depth complexity sampling	✓	✓	✓	rect/sph
(4) Analytic solution				
4.4.3 Convolution textures	✓	✓		polygon
4.4.3 Occlusion textures	✓		✓	polygon
4.7.2 Cone tracing	✓	✓	✓	sphere
4.7.2 Beam tracing	✓	✓	✓	polygon

Table 4.1. Soft shadow algorithms.

the above algorithms account for outer penumbrae and all dynamic ranges. Finally, a "maybe" in the above table indicates that it is possible to achieve/avoid the feature/artifact given a specific implementation.

In the next chapter, we present literature on other treatments of shadows, including soft shadows as a result of ambient occlusion, precomputed radiance transfer, global illumination, and motion blur.

CHAPTER 5

Other Treatments of Shadows

5.1 What's Covered in This Chapter

This chapter covers other treatments of shadows. The topics include

- Bump-mapped surfaces taking self-shadowing into proper account (Section 5.2) [573].

- Treatments of masking and self-shadowing within advanced reflection models (Section 5.3).

- "Quasi" shadows from semitransparent and translucent surfaces (Section 5.4). It is quasi because these are technically not creating shadows. They use shadow algorithms not to compute a region not occluded from light, but a region that represents the transmissivity of light through the semitransparent or translucent surfaces.

- Highly complex thin materials, such as hair and fur (Section 5.5).

- Atmospheric shadows to simulate fog-like environments taking into account occlusion of light (Section 5.6). Though this is better known as "volumetric shadows" in the literature, we hesitate to call it that only because it can be mistaken for shadows as applied to volume graphics (voxels).

- Soft shadows as a result of motion blur, which are critical for film and video output (Section 5.7).

- Soft shadows as a result of ambient occlusion, which fake sky light with the look of soft shadows as if all objects are under an overcast day (Section 5.8) [7, 403].

- Soft shadows as a result of precomputed radiance transfer, which use spherical harmonics and other forms of hierarchical directional representations in diffuse, low-frequency lighting environments (Section 5.9) [542, 203, 474].

- Soft shadows as a result of global illumination, which includes discussions on radiosity-based, instant-radiosity-based, and Monte-Carlo-based ray tracing techniques (Section 5.10) [103, 530, 273, 384, 271, 610, 526, 151].

5.2   Bump Mapping

As indicated in Section 1.4.2, bump mapping by itself does not take the self-shadowing effects into account, thus making the bump-mapped surface appear somewhat flat looking. In this section, we review the techniques used to provide self-shadowing for bump maps.

Horizon mapping [391] approximates the shadows cast by the bumps on the same surface. Interpreting the bump function as a 2D table of height values, horizon mapping computes and stores, for each point on the surface, the angle between the horizon and the surface plane at eight or more azimuthal directions on the surface plane (see Figure 5.1). During rendering, the horizon angle at the intersection point is interpolated from the light direction and the horizon map. If the horizon angle from the surface tangent exceeds the angle to the light, then this point lies in shadow. The horizon map assumes that the surface on which the bump map is applied is planar. Max [391] also introduces an approximate correction factor, to take into account some moderate deformations caused by this underlying surface.

Sloan and Cohen [538] extend the horizon-mapping algorithm for the GPU using multipass rendering of texture lookups and stencil buffering (Figure 5.2). The eight horizon mapping directions are encoded into two four-channel texture maps and multiplied with the basis functions (expressed as a texture map) to determine the amount of shadowing. Note that this algorithm has been implemented and is readily available on DirectX. Kautz et al. [294, 238] also permit self-shadows from bump maps on the GPU. As a preprocess, rays are shot into the upper hemisphere for each bump-map pixel in many directions, and a tight ellipse that contains as many unobstructed rays as possible is computed (see Figure 5.3). To determine shadowing per pixel, the lighting direction must have a hit inside this ellipse to be

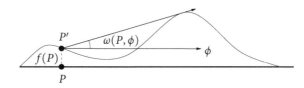

Figure 5.1. Horizon angle at P as a function of the azimuthal direction.

Figure 5.2. Bump-mapped surface without (left) and with self-shadowing (center and right). *Image courtesy of Sloan and Cohen [538].*

lit; otherwise, it is in shadow. Testing to determine if it is inside of the ellipse is done by fast dot products using the GPU.

To improve on the Sloan and Cohen [538] performance, Forsyth [179] avoids the need to do multiple render passes for real-time horizon mapping by storing the multiple horizon maps in a 3D hardware texture. Onoue et al. [439] then introduce a distance map (in addition to the horizon map) to improve the accuracy of shadows when the bump map is applied to high curvature surfaces.

Note that horizon mapped shadows are also useful in shadow determination for heightfields (see Section 3.8). However, because the self-shadowing for bump maps can be approximate, especially when high-frequency changes are not common, shadowing for heightfields need to be extended for better accuracy.

The above approaches compute self-shadowing of bump-mapped surfaces onto themselves. Noma and Sumi [435] use ray tracing to cast shadows from other surfaces onto a bump-mapped surface. This is done simply by offsetting the bump-mapped surface as the starting shadow-ray origin. Shooting the shadow ray with this offset can make convincing shadows on the bump-mapped surface. For complex bump maps though, since the offset is computed locally at the intersection point, the offset will neglect bumps further away.

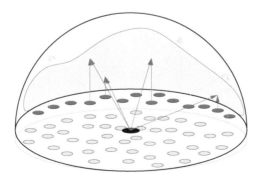

Figure 5.3. Tight ellipse (in green) containing unobstructed rays.

5.2.1   Advanced Bump Mapping Effects

Though bump mapping is widely used, it is limited in that the silhouette of the surface is not "bumped," which is very evident when the surface is viewed from up close or when the shadow of the surface is up close. Also, bump mapping does not account for proper adjustments when viewing from different angles.

The latter issue is improved upon by applying parallax mapping [312] and its variations, where the (u, v) texture coordinates are modified to allow proper adjustments when viewing from different angles. However, no attempt at proper shadowing has been done to date.

As for the "bumped" silhouette, displacement mapping [110] is introduced where the surface is actually displaced (and retessellated in some form). Due to the details that must be captured properly, shadow computations are trivially accounted for using any shadow depth map or ray tracing approaches; however, such surface displacements tend to require significantly more retessellated triangles, and thus the performance of displacement mapping tends to be slower, and this technique is often used in offline rendering scenarios.

To permit faster bumped silhouettes with proper parallax computations and shadowing (including self-shadowing and shadowing of/from other objects), two directions have surfaced without the need for extreme tessellation to represent the displacement:

- On-the-fly computations of the exact hit points with the displacement [396, 626, 581, 582], bounded locally by the original surface to its maximum perturbed height (denoted by h). The main examples of such techniques, which also consider shadowing, include relief mapping [626], steep par-

Figure 5.4. Traditional bump mapping (left) versus relief mapping (right). Note the proper shadows on the mapped surface and the floor. ©*2005, Advanced Game Development with Programmable Graphics Hardware, by Alan Watt and Fabio Policarpo. Reproduced by permission of Taylor and Francis Group, LLC, a division of Informa Plc.*

allax mapping [396], parallax occlusion mapping [581, 582], and cone step mapping [85]. These techniques are similar, but differ by the way hit points are computed. Watt and Policarpo [626] use iterative approaches (linear and binary searches) to converge to a hit point within h, McGuire and McGuire [396] use ray voxel traversal within h, Tatarchuk [581, 582] uses a linear search towards a piecewise linear curve approximating the perturbed height, and Chang et al. [85] use visibility cones to intersect against the area light to determine the amount of occlusion. See Figure 5.4 for a sample rendering of relief mapping.

- Precomputed visibility to allow very fast exact hit points to be computed, such as view-dependent displacement-mapping techniques [619, 620], where a 5D function (grid) is used to store precomputed visibility due to the displacement. The rendering is performed on the original nondisplaced geometry and on a combination of this 5D function.

5.2.2   Trends and Analysis

Bump mapping itself has long been used in film, video, and games. However, self-shadowing computations for bump mapping have not been that widely adopted, likely because if the bump is important enough to account for self-shadowing effects, it is important enough to need to take into account the bumped silhouette and its shadow implications, which bump mapping cannot do. When shadowing is needed for bumps, typically, relief mapping or displacement mapping are used instead—relief mapping typically for real-time needs and displacement mapping for offline renderings.

The study for self-shadowing of bump maps remains important because the main technique, horizon mapping, has become an interesting research direction for rendering heightfields with shadows (Section 3.8) as well as ambient occlusion (Section 5.8).

Also note that the term mesostructure has also been used in the literature to describe the detailed features of a surface, such as bump mapping, relief mapping, displacement mapping, etc.

5.3   Advanced Reflection Models

As indicated in Section 1.4.3, advanced reflection models simulate how light reflects off a surface, assuming that an infinity of surface details are integrated at the point to be shaded. This phenomenon represents a special case in which proper self-shadowing computations are needed to achieve the correct visual appearance, even though individual shadows are not visible. The visual impact of self-shadowing is often more amplified at grazing angles for the viewing or lighting directions. Note, however, that in most reflection models, these surface details are considered

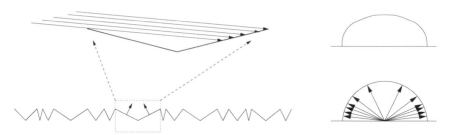

Figure 5.5. Masking and self-shadowing in the V-shaped microfacets model of Torrance and Sparrow [589]. The distribution of surface normals at the bottom right can be approximated to a shape at the top right, providing the fraction of surface normals oriented in each direction.

much larger than the wavelength of light, thus limiting themselves to geometrical optics. In this section, we go over the details of such shadowing computations for advanced reflection models.

In a reflection model proposed by Torrance and Sparrow [589], the surface details are represented by a collection of mirror-like microfacets randomly oriented, thus simulating isotropic reflection. Assuming that the microfacets are small but elongated symmetric, V-shaped grooves, as illustrated in Figure 5.5, and that both the lighting and the visibility can be considered at infinity (i.e., directional), the masking and self-shadowing effects on each pair of V-shapes can be computed analytically, resulting in visible portions of the grooves that are either illuminated or in shadow. The appearance of roughness in the reflection model is captured by a Gaussian distribution of the openings of these V-shapes. Trowbridge and Reitz [592] represent the microfacet distribution from ellipsoids of revolution, and their formulation was later adopted by Blinn [53] in a commonly used reflection model in computer graphics. While different microfacet distributions have been introduced, they very often follow similar shapes. Several reflection models are based on this microfacet theory. Using the same distribution of V-shapes but replacing the mirrors with perfectly diffuse reflectors, Oren and Nayar [440] show how diffuse interreflections (radiosity) affect the appearance of rough surfaces. Their incident illumination also assumes microfacet self-occlusions.

Cook and Torrance [109] consider the spectral composition of incident light and reflection, in order to simulate the color-shift specular effect due to the Fresnel factor. Their reflection model achieves a very realistic appearance of metallic surfaces.

In their Beard-Maxwell model, Maxwell et al. [392] simulate the appearance of painted surfaces, with surfacic and volumetric reflections. A distribution of microfacets is assumed, but its self-shadowing formulation is different. It is based on a function of two free parameters that can be associated to empirical observations or user-determined in order to control the fall-off of reflection.

Smith [543] describes the surface as a Gaussian distribution of heights, and formulates the probability that a point of a surface at a given height and of given slopes (in x and y on the surface plane) be in self-shadow, given an incident angle of illumination. His formulae fit well with simulated experimental results.

He et al. [230] use Smith's model [543] as a multiplicative shadowing factor and modify the probability of shadowing and masking to take into consideration the effective roughness, i.e., the roughness according to the illumination and viewing angles, which can be very different at grazing angles than the actual surface roughness. The physically based reflection model is consistent with its various terms and it handles polarization. It has shown good comparisons with measurements of different types of materials, including metals, nonmetals, and plastics, with smooth and rough surfaces. As a result, it forms, even to this day, the most complete analytical reflection model in computer graphics.

Ashikhmin et al. [22] introduce a very general microfacet distribution based on normal distribution, therefore, not limited to heightfields, and from which only first-bounce reflection is considered. They observe that the microfacet distribution has the largest visual impact on mostly specular reflection surfaces and consider a simple shadowing term to be sufficient as long as the reflection remains physically plausible. Their shadowing term assumes uncorrelated microfacets, and therefore, they simply multiply the two probabilities associated with the viewing and lighting directions as independent functions. Because this can underestimate reflection intensity, in order to include a form of correlation, they introduce a linear interpolation based on the orientation away from the surface normal between the minimum of both terms (correlated) and the product of both terms (uncorrelated). The reflection models thus generated are very flexible, and examples demonstrate the quality of the approximation in comparison with several different specialized reflection models.

However, for many types of surfaces, the microfacets have preferred orientations. The shadowing effect then also depends on the orientation of the microfacets relative to the surface. In the case of simple distributions of microfacets, the self-shadowing can be computed analytically, such as in the work from Poulin and Fournier [466], where they analytically compute the masking and self-shadowing caused by adjacent parallel cylinders simulating an anisotropic surface (Figure 5.6).

While most reflection models assume the ray theory of light, Stam [556] uses the wave theory of light in order to simulate related phenomena, such as diffraction. He extends the model of He et al. [230] to handle anisotropic reflection with a correlation function of a Gaussian random surface and uses a shadowing term accordingly.

Instead of using statistical models to represent the microstructure of a surface, Cabral et al. [77] define a heightfield as a polygonal mesh and compute the direct illumination over this surface for all combinations of light-camera orientations. The result is an anisotropic reflection expressed as a 4D table. In order to speed up shadowing computations, they rely on horizon mapping [391] over the heightfield.

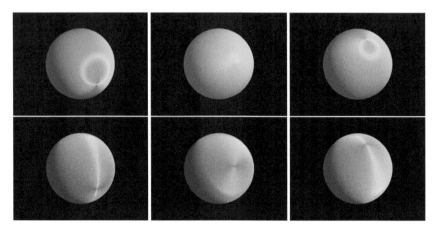

Figure 5.6. Anisotropic reflection model off a rotating sphere. Longitudinal scratches (top) and latitudinal scratches (bottom).

The method of Heidrich et al. [238] for shadowing bump-mapped surfaces (described in Section 5.2) can also be used to efficiently compute shadowing in the microgeometry of BRDFs or BTFs simulated by heightfields or bump maps.

A common usage of anisotropic reflection models comes from the modeling of hair, in which proper lighting and shadowing need to be considered [313, 303]. Hair also has the property of lots of small details, in which the approaches for proper shadowing are handled very similarly to semitransparency (see Section 5.4.2).

5.3.1   Trends and Analysis

The vast majority of reflection models used in image synthesis for applications ranging from film production to real-time video games are simplified models associated with early work of Phong [462] and Blinn [53]. Their popularity is partly due to their efficiency and the ease of control by artists, as well as to code legacy. Standard APIs such as Direct3D and OpenGL continue to support Phong's model.

The years of advances in reproducing realistic appearances, whether by capturing reality from measurements or by modeling the behavior of various aspects of reflectance, have resulted in a broad set of advanced reflection models and BRDF/BTF databases publicly available. Such reflection models are often readily included as advanced "shaders" in commercial rendering systems or offered as "plug-in" or specialized libraries.

The flexibility of real-time shaders on the GPU allows most of these advanced reflection models to be used, with potentially some impact on rendering speed depending on the complexity of the shader.

Advanced reflection models are more common in predictive rendering and in applications requiring accurate appearance simulation, such as in the car industry, in lighting and material design, and in architecture.

5.4 Semitransparent Surfaces

Recall in Section 1.2.3 that colored shadows are present when semitransparent surfaces are involved. There are three basic techniques to render shadows from semitransparent surfaces: ray tracing (Section 5.4.1), shadow depth maps (Section 5.4.2), and shadow volumes (Section 5.4.3). Also note that the majority of the shadow algorithms for volume graphics (Section 3.7) support semitransparent shadowing.

5.4.1 Ray Tracing

The earliest shadow algorithms to support semitransparent surfaces come from ray tracing approaches. Hall and Greenberg [221], Lee et al. [340], and Pearce [454] perform the usual shadow-ray computations but do not stop at the first semitransparent occluder, and they attenuate light as a function of the angle between the shadow ray and the normal at the intersection between the shadow ray and any object the shadow ray hits. They also need to ensure that the occluder-shadow-hits are sorted properly for the attenuation computations. See an example of the shadow attenuation effect achieved in Figure 5.7.

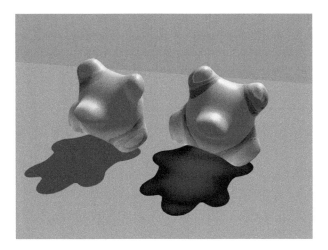

Figure 5.7. Semitransparent shadows from ray tracing, using no shadow attenuation on the left, and full shadow attenuation based on normals on the right. *Image courtesy of Lee Lanier.*

The above algorithms cannot take into account the refraction that takes place. Shinya et al. [524] apply fundamentals of paraxial approximation theory to achieve semitransparency, concentration, and diffusion of light (in the form of semitransparent shadows). A pencil is defined by the axial ray surrounded by nearby paraxial rays. For semitransparent shadows, pencils are shot from the lights to the bounding volumes of all semitransparent objects, then propagated to the surfaces. Thus, refraction is taken properly into account, but only first-generation transmitted light is taken into account.

For faster computations, the shadow depth map and shadow volume variations (to be discussed in the next sections) have been extended to account for such shadows, though no attenuation or refraction of light is achieved.

5.4.2   Shadow Maps

Lokovic and Veach [362] introduce deep shadow maps, in which a shadow map pixel contains a 1D function of (R,G,B) transmittance and depth information, sorted in depth order (see Figure 5.8). Each point to be shaded accesses its shadow map pixel, determines where it resides with respect to the transmittance function, and returns a shadow result. Because the transmittance function carries a lot more information per shadow depth pixel, the authors indicate that a lower-resolution deep shadow map can match the quality (i.e., capture the details well) of a much higher-resolution regular shadow depth map. Salvi et al. [495] attempt deep shadow maps on the GPU. Note that the deep shadow map is also capable of dealing with soft shadows (Section 4.5.4), motion blur (Section 5.7), and highly complex thin materials (Section 5.5), and is supported in volume graphics (Section 3.7.5). See Figure 5.9 for a result with semitransparent colored shadows, and a result with shadows with motion blur.

Dachsbacher and Stamminger [115] use a structure based on the shadow depth map to store depth, surface normal, and irradiance (incoming illumination) of the directly illuminated points of a translucent object. To efficiently determine the sub-

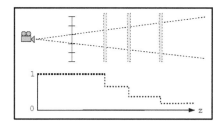 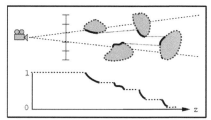

Figure 5.8. A 1D function of (R,G,B) transmittance and depth information, due to semi-transparency (left) and highly complex thin materials (right). ©2000 ACM, Inc. Included here by permission [362].

Figure 5.9. Semitransparent shadows from deep shadow maps, and with motion blur. *Images rendered by Victor Yudin in 3Delight. Images courtesy of DNA Research.*

surface light contribution at a point to be shaded, the point is projected into the translucent shadow map, and the values within a determined shadow map radius are hierarchically (mipmap) filtered with a pattern of 21 samples according to a formulation for the translucency computation. The current technique is, however, limited to translucent objects for which all visible points can be linked by a line segment within the translucent material to the contributing points directly illuminated by the light source.

Xie et al. [649] ray trace multilayer shadow maps (Section 4.7.4) and include an opacity value with the depth value so that semitransparent shadows and soft shadows can be rendered.

Enderton et al. [159] apply a random screen-door mask for each semitransparent triangle to a regular shadow depth map and refer to it as a "stochastic transparency shadow map." This allows processing to be faster than generating an absolutely detailed result, but it can produce stochastic noise, of which filtering is applied to smooth out the results. McGuire and Enderton [394] extend the above

Figure 5.10. Regular shadow depth map versus stochastic transparent shadow map. ©2011 *ACM, Inc. Included here by permission [394].*

approach to allow for color to be properly transmitted, even with multiple intersections of the shadows from semitransparent objects. See Figure 5.10 for an illustration of a stochastic transparency shadow map.

More shadow maps supporting semitransparency are available that are more well suited to hair and fur and can be seen in Section 5.5. The two subjects (semitransparent material and hair/fur) are actually quite similar in certain cases. Further, semitransparent surfaces also contribute to atmospheric shadows, a topic discussed in Section 5.6, though only covered by McGuire and Enderton [394].

5.4.3   Shadow Volumes

Kim et al. [306] compute shadows from semitransparent surfaces by modifying the shadow count from incrementing/decrementing by 1 to incrementing/decrementing by the transparency multiplier. This modification makes the algorithm difficult to implement on the GPU. Also, due to lack of depth information, correct self-shadowing for semitransparent surfaces cannot be guaranteed, and the transparency attenuation of the shadows cannot be guaranteed.

Haselgren et al. [228] support transparency mapping to arrive at hard, colored shadows. For each semitransparent triangle, shadow volumes are generated, and an extra render pass is needed. This approach is useful when there are few semitransparent triangles, such as the complex netting in Figure 5.11 represented by a few transparency-mapped triangles. Forest et al. [177] combine the above approach with penumbra wedges (instead of shadow volumes) to achieve the soft shadow equivalent.

Figure 5.11. Soft shadows from transparency mapping. Image courtesy of Forest et al. [177]. Computer Graphics Forum ©2009 The Eurographics Association and Blackwell Publishing Ltd. Published by Blackwell Publishing, 9600 Garsington Road, Oxford OX4 2DQ, UK and 350 Main Street, Malden, MA 02148, USA. Reproduced by permission of the Eurographics Association and Blackwell Publishing Ltd.

Sintorn et al. [536] use per-triangle shadow volumes to robustly support both semitransparent and transparency-mapped shadows. See Section 2.4.2 for further details.

Keep in mind that while supporting semitransparency, many of the optimizations discussed in Section 2.4.3 cannot be applied as they are, as semitransparent shadows will require some more information in addition to the opaque shadowing situation.

5.4.4   Trends and Analysis

Semitransparent shadows are important in offline renderings. Deep shadow maps are the norm in film and video rendering solutions, with ray tracing variants as the backup solution. In CAD offline renderings, usually only ray tracing solutions are offered.

Semitransparent shadows are not as commonplace for real-time engines, though variants of the stochastic transparency and the per-triangle shadow volumes show promise.

5.5   Highly Complex Thin Materials

For highly complex thin materials such as hair and fur, extreme care must be taken to ensure good-quality rendering results to handle both self-shadowing and shadowing onto other objects. Due to the complexity of such materials, shadow volume techniques have not been offered in this domain. Also due to the need to capture the tiny features, raw ray tracing techniques are also not recommended. Thus, we are mostly left with shadow map–based techniques and volume-graphics techniques to date, the latter of which has already been covered in Section 3.7.

We have also seen some shadow map–based techniques able to support highly complex thin materials from previous sections, such as deep shadow maps [362] and ray tracing multilayer shadow maps [649] from Section 5.4.2, as well as alias-free shadow maps [4, 275] from Section 2.3.7.

Early attempts at computing shadows for hair and fur include research from the following papers [339, 313, 303]. The major trend since then has revolved around opacity shadow maps [304], which can be considered slice-based approximations to deep shadow maps [362]. The opacity shadow map approach creates slices of opacity-map planes, each perpendicular to the light source. For each slice plane, the opacity of each pixel is determined for the region that crosses the opacity map plane and encloses the opacity pixel. During shading, the point to be shaded is projected toward the slice planes and their corresponding pixels, and the shadow occlusion is derived from the pixels' opacity, including interpolation between the slice planes using GPU alpha blending. Figure 5.12(left) shows the slice planes (in green) and the pixels needed for consideration for a particular ray.

Figure 5.12. Opacity map of uniformly spaced slice planes (left) versus deep opacity map, which wraps the slice planes around the hair structure (right). *Image courtesy of Yuksel and Keyser [655]. Computer Graphics Forum ©2008 The Eurographics Association and Blackwell Publishing Ltd. Published by Blackwell Publishing, 9600 Garsington Road, Oxford OX4 2DQ, UK and 350 Main Street, Malden, MA 02148, USA. Reproduced by permission of the Eurographics Association and Blackwell Publishing Ltd.*

In the original implementation [304], the slice planes are placed uniformly, which requires many such planes to achieve good results. Some generic techniques have been pursued to get better performance, such as

- More extensive use of the GPU towards faster opacity mapping performance via simplified geometric assumptions of hair [316], multiple render targets (rendering 16 slice planes in one render pass) [424], and rendering in depth buckets [533].

- Better placement of the slice planes, such as adaptive density clustering [405], and wrapping the slice planes around the hair structure [655], a technique called deep opacity maps (see the green lines in Figure 5.12(right)). A rendering comparison of deep opacity maps versus opacity maps can be seen in Figure 5.13, which is also a good indication of the state-of-the-art capabilities of (in this example) hair rendering.

- Representing the hair geometry in a highly compressed, voxel format instead of slice planes [47, 155, 534].

- Reconstructing the transmittance function along each ray using a Fourier series and additive blending as Jansen and Bavoil [265] do. The coefficients of the Fourier series are stored in the so-called Fourier opacity map. This provides memory savings as well as smoother results than slice-based solutions.

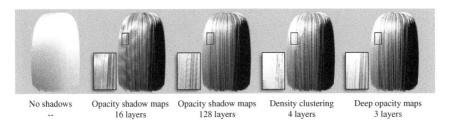

No shadows	Opacity shadow maps	Opacity shadow maps	Density clustering	Deep opacity maps
--	16 layers	128 layers	4 layers	3 layers

Figure 5.13. No shadows versus opacity map and deep opacity map. *Image courtesy of Yuksel and Keyser [655]. Computer Graphics Forum ©2008 The Eurographics Association and Blackwell Publishing Ltd. Published by Blackwell Publishing, 9600 Garsington Road, Oxford OX4 2DQ, UK and 350 Main Street, Malden, MA 02148, USA. Reproduced by permission of the Eurographics Association and Blackwell Publishing Ltd.*

5.5.1 Trends and Analysis

Due to the complexity of such highly complex thin materials, shadow volume techniques have not been offered in this domain. Also, due to the need to capture tiny features, raw ray tracing techniques are also not recommended, unless one is prepared to accept extremely high sampling densities.

In offline rendering situations, the deep shadow map remains the most convenient implementation, and ray tracing multilayer shadow maps [649] have been used successfully in production. In real-time rendering situations, some variation of the opacity mapping seems to be the logical choice, but it is rare to see such capabilities in real-time.

5.6 Atmospheric Shadows

Sunlight scattering in the air causes the atmosphere to glow. This glow is particularly visible in the presence of shadows. As usual, a ray is shot to determine the closest visible surface, and the critical question is not just if the intersection point is in shadow. The segments along the ray (not shadow ray) that are visible from the light are just as crucial (see Figure 5.14). The information is necessary to acquire atmospheric shadows assuming only a single scattering model for light diffusion. Single scattering assumes that incoming light from the light source reaches a point along the viewing ray and deviates only once in viewing direction. If the ray is not illuminated, then the shading calculations are the same as in the original illumination model, including direct attenuation due to the media. However, with partial illumination, an additional component (atmospheric shadows) is included. In the literature, these techniques are also often referred to as volumetric shadows.

In the subsequent sections, we discuss the three common techniques used to achieve atmospheric shadows: shadow volumes (Section 5.6.1), shadow depth maps (Section 5.6.2), and global illumination (Section 5.6.3).

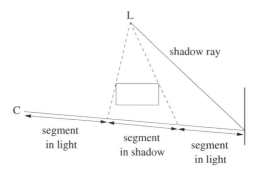

Figure 5.14.   Shadowing in the presence of light scattering media due to the above illuminated/not-illuminated ray segments.

5.6.1   Shadow Volumes

For each pixel, Max [390] and Nishita et al. [431] use shadow volumes to calculate the exact ray segments that are visible from the light to achieve atmospheric shadows, employing a single scattering model. This approach can be extended to voxel occlusion testing (Section 2.5.1), in which the voxel occlusion values that the viewing ray traverses are used to determine which regions illuminate or shadow the ray segment. The span of the ray segment within each voxel can be trivially computed since it is readily available in voxel traversal algorithms [9]. Similarly, Ebert and Parent [152] also use subdivision grids to accelerate atmospheric shadows.

The above approaches are all CPU-based. James [264] and Biri et al. [51] apply shadow volumes on the GPU where the overlapping volumes are accumulated using frame-buffer blending with depth peeling.

Wyman [648] voxelizes the shadow volumes and conceptually does a 1-bit version of voxel occlusion testing (the voxel is either lit or shadowed). However, voxelization occurs in epipolar space, and traversal of the viewing ray can require as little as a texture lookup on the GPU. Impressive performance numbers are reported for this approach, though there may be bottlenecks residing in the voxelization step because a fairly high voxel resolution is needed to generate good results.

Billeter et al. [49] suggest that a shadow depth map already defines a volume enclosing the directly illuminated space. Thus, a mesh surrounding this space can be created and rendered in real-time with an accurate attenuation without the need for traversing the shadow depth map per sample to determine visibility and attenuation. However, the complexity of this approach is huge when the resolution of the shadow depth map becomes large and problematic when the resolution is insufficient.

5.6.2   Shadow Depth Maps

It is possible to achieve atmospheric shadows from shadow depth maps as well, assuming a conical atmospheric effect from a spotlight. Instead of comparing depth values from the point to be shaded, the viewing ray is projected to the shadow depth map pixels, and depth comparisons are done per shadow depth map pixel (see Figure 5.15). An accumulation of the shadow tests determines the amount of shadowing that occurs in the atmospheric effect. However, doing the comparison per shadow depth map pixel can be quite expensive because the viewing ray can potentially extend through the entire shadow depth map. To improve speed, only every (stochastically chosen) nth shadow depth map pixel comparison is done; this, however, can result in noise in the atmospheric shadows. Also note that in order for the above approach to work, the Z-depth values stored must be the smallest Z-depth value; thus, alternative Z-depth values and surface IDs cannot be used for avoiding self-shadowing (Section 2.3.4).

The above approach is often referred to as ray marching. However, ray marching can also mean that at each point to determine light visibility, ray tracing (shadow rays) can be used instead of sampling shadow depth map pixels, though ray tracing tends to be more expensive. Though the above approach has been known and well implemented in offline rendering solutions since the early 1990s, the real first mention in publication is seen in the paper of Gautron et al. [186].

To achieve the above approach on the GPU, fixed sampling planes parallel to the viewing direction are used to record atmospheric shadow/visibility from the shadow depth map information [134]. However, this approach can easily result in banding artifacts for the atmospheric shadows. The use of subplanes between the sampling planes are introduced by Dobashi et al. [135], where the need for subplanes is determined by the intensity of the region between the sampling planes. In this way, the banding artifacts can be reduced without need for a significantly higher number of sampling planes. Mitchell [412] goes over some implementation details of the sampling plane approach using GPU alpha-blended planes,

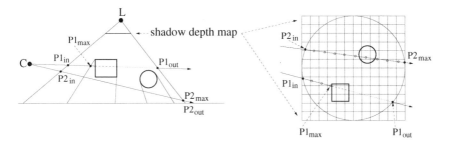

Figure 5.15. Ray marching using shadow depth maps.

Figure 5.16. Combination of colored semitransparent shadows and colored atmospheric shadows. ©2011 ACM, Inc. Included here by permission [394].

while McGuire and Enderton [394] extend Mitchell's approach, taking into account attenuation from semitransparent surfaces (see Figure 5.16). Imagire et al. [257] reduce the number of sampling planes by averaging the illumination over regions near each plane. In general, these approaches require expensive preprocessing and high fill rates.

To reduce the frequency of ray marching in shadowed areas, Wyman and Ramsey [647] use shadow volumes. On a per-pixel basis, the absence of shadow volumes means no ray marching is needed; the presence of shadow volumes indicates regions requiring ray marching. Toth and Umenhoffer [590] avoid the entire ray-marching operation for certain samples and use interleaved sampling to borrow results from neighboring samples. Engelhardt and Dachsbacher [162] use epipolar sampling so they need only ray marching depth discontinuities along the image-space epipolar lines. Baran et al. [34] use a partial sum tree to indicate state changes with respect to the visibility of the ray segments within a canonical configuration, i.e., an orthographic view and parallel lighting situation. Under such situations, the lack of state changes indicates that no ray marching is needed, as the previous ray is identical. Epipolar rectification of the shadow depth map then permits a general view (e.g., perspective) and lighting (e.g., spot or point light) configurations. Chen et al. [90] extend Baran's approach to the GPU.

5.6.3   Global Illumination

Rushmeier and Torrance [492] apply zonal methods used in heat transfer to the computation of radiosity (Section 5.10.1). Their method discretizes the medium into small volumes for which the form factors volume/volume and volume/surface are calculated. The shadows are generated with the hemicube extended to a full cube.

5.6.4   Trends and Analysis

In the computation of atmospheric shadows, both shadow depth maps and shadow volumes have their own merits. The choice for hard shadows will likely match the choice for atmospheric shadows due to the ease of code maintenance as well as the ability to avoid any mismatches in the shadow computations, i.e., if shadow volumes are used for hard shadows in the current application, it makes sense to choose the shadow volume approach for atmospheric shadows as well.

The shadow depth map's CPU ray-marching variation seems to be the most used in industry for offline rendering solutions. This is likely because the shadow depth map is already implemented in the same offline rendering solution, so this capability is an extension. Atmospheric shadows have been seen in film quite often nowadays.

For real-time purposes, Mitchell [413] provides a fast but approximate post-processing approach. Among the shadow depth map variations, the approaches by Baran et al. [34] and Chen et al. [90] seem the most promising. Among the shadow volume variations, the approach by Wyman [648] seems the most promising. However, adoption of the choice of algorithms is unclear at this point.

Note that some of the occlusion culling techniques as discussed in Section 2.3.2 cannot be used as they are, due to the need to account for invisible shadow casters that are visible atmospheric shadow casters. Similarly, shadow depth map variations addressing perspective aliasing (Section 2.3.7) may no longer be appropriate, because the effect of the atmospheric shadows cannot be easily localized to determine ideal resolution. Also note that atmospheric shadows are used more often in conjunction with hard shadows. This is because if soft shadows were involved, not only would the performance be significantly slower, the visual effect of the atmospheric shadows would be less pronounced.

5.7   Motion Blur

Motion blur is now used extensively in rendering images for film and video. Without motion blur, the renderings appear noticeably jumpy in the temporal domain over the animated frames (e.g., in stop-motion filming techniques [366]) and, as a result, not as convincing as a special effect. With respect to its shadows, the blurred

Figure 5.17. Shadows cast with motion blur. *Image rendered by Victor Yudin in 3Delight. Image courtesy of DNA Research.*

(thus soft) shadows come from the occluder's or light's motion captured in a single frame. See Figure 5.17 for a rendering of motion blur included.

Other than distribution ray tracing [108] and its optimized variants (see Section 4.7.1), which shoot rays that intersect objects in stochastically chosen time domains, there has been little published research on achieving motion-blurred shadows. It may sound like a simple thing, but it is not, though it is simpler in a ray tracing environment. The problem is complex enough when just considering motion blurring of hard shadows (resulting in soft shadows) from a moving occluder but stationary receiver and light, but it gets even more complex when motion-blurred soft shadows (from extended lights) need to be considered, or motion-blurred shadows from a moving light.

One can render motion-blurred shadows by pregenerating several shadow depth maps, each shadow depth map indicating a different snapshot in time. This can be followed by integrating the shadowing based on sampling each of the shadow depth maps. However, the merging of such results likely produces banding of the blurred shadows if not enough time snapshots are done, which is very visually distracting in an animation. Filtering the banding results may cause shifts over an animation. This banding can be reduced by stochastically shifting each frame's rendered images by a very small amount [548] (though the authors apply the approach over interlaced fields for video animations).

One could also do a time-jitter per shadow depth map pixel. However, because each shadow depth map pixel may project to multiple screen pixels, the motion blurred shadows look very blocky. Within the concept of deep shadow maps, Lokovic and Veach [362] apply this trick to the transmittance functions of the shadow map pixel; this is essentially jittering in 3D and thus can achieve decent motion-blurred shadows.

As mentioned before, the complexity of the problem is greatly increased if such variables as moving occluders, moving lights, and extended lights are considered. Sung et al. [569] formalize the motion-blur equation without explicitly considering shadows:

$$i(\omega, t) = \sum \int \int R(\omega, t) g(\omega, t) L(\omega, t) \, dt \, d\omega,$$

where ω is the solid angle intersected with a pixel at time t, $g(\omega, t)$ is the visibility of the geometry, $R(\omega, t)$ is the reconstruction filter, and $L(\omega, t)$ is the shading equation. Imagine the worst-case scenario in which both $g(\omega, t)$ and $L(\omega, t)$ must be folded into the penumbra integrand in order to compute a motion blurred soft shadow; the above integral becomes very difficult to solve. Actually, that is even not exactly correct either, because the $g(\omega, t)$ term needs to be part of the soft shadow integral.

5.7.1   Trends and Analysis

There are basically two feasible approaches to accounting for motion-blurred shadows: distribution ray tracing [108] and deep shadow maps [362]. Both approaches are very flexible, and both have been successfully used in film and video, though distribution ray tracing tends to be slower. While motion blur is available for real-time applications, it is rare to see motion-blurred shadows in that domain.

5.8   Ambient Occlusion

Ambient occlusion is the term coined to capture the darkening effect of more enclosed (or occluded) regions when the incident illumination comes from all its surrounding. In the presence of ambient occlusion, creases and holes appear darker than more-exposed areas, and contact shadows of objects resting on a surface look more plausible. Ambient occlusion results in softer (lower frequency) shadows due to outdoor skylight over diffuse surfaces, and thus provides visual clues about depth, curvature, and proximity of elements around a surface and between neighboring objects in a 3D scene.

Because of its appearance, ambient occlusion is often thought of as a simple replacement for computationally intensive soft indirect illumination or for overcast sky or dome illumination. The constant "ambient" term used in a local reflection model results in a constant-value shading. The ambient occlusion term adds variations that increases realism and therefore has been quite popular since its introduction. Not surprisingly, some form of ambient occlusion shading appears in most commercial rendering packages, as well as in more and more game engines.

5.8.1   Related Concepts

In some of the earliest work to achieve skylight and shadowing, Nishita and Nakamae [429] generate a hemisphere representing the sky and band sources are defined
longitudinally on this hemisphere. By assuming a uniform incident illumination
of the sky within each band, they can sample the luminance over the centerline of
each band. The luminance of the sky can be interpolated from clear to overcast.
Every object in the scene is projected onto the sample lines in order to compute
the occlusion factor. If the point to be shaded is on a tilted surface, the hemisphere
is also tilted by the same angle in which the light reflection from the ground is
considered for this tilted angle.

Miller [410] introduces the notion of accessibility as the radius of a spherical
probe that can lie at a point on a surface. It can be evaluated by growing a sphere at
the point until it touches another polygon, or by shooting rays and keeping the closest one. This accessibility value can be used to determine how much dust or other
material can accumulate in a region, or how sandpaper cannot polish (remove)
excess material. Ambient occlusion corresponds to accessibility of ambient light.

Stewart [564] integrates his vicinity shading to improve the perception of isosurfaces computed from volumetric data. His method estimates the vicinity for a
set of rays traversing voxels in the hemisphere above a given voxel. Stewart greatly
optimizes the computations by exploiting the regularity of voxels and the space
of global fixed directions traversing the set of voxels. He also assumes that only a
voxel of larger density completely occludes a sampled direction issued at a given
voxel. This allows him to efficiently compute in preprocessing the vicinity shading
for all voxels for any hemisphere radius, and for all possible values of isosurfaces.

Hernell et al. [242] include semitransparency of voxels in the evaluation of the
ambient occlusion factor, and improve on the computations by exploiting adaptive
sampling, multiresolution optimizations, and GPU.

A number of authors extend the notion of ambient occlusion in order to include
light intensity, shading, and local light's multiple interactions with reflections and
refractions. They sometimes refer to their methods as "obscurances" to distinguish
from ambient occlusion. However, with more and more papers mixing the concepts of obscurances and ambient occlusion, we will simply use ambient occlusion
in the following. In any way, the evaluation of obscurance requires a form of local
visibility, which is inherited from ambient occlusion methods. Because we deal in
this book mostly with shadowing and not shading, we refer the reader to the survey
of Mendez-Feliu and Sbert [403] and a number of their original publications.

5.8.2   Representation

Even though ambient occlusion appeared in one form or another (Section 5.8.1 and
[674, 403]), it is mainly after presentations in the early 2000 [332, 98] that this concept has taken over as the dominant approach of realistically generating a skylight

look in the film and game industries. Figure 5.18 gives an example rendering using ambient occlusion.

The basic idea is to compute, for each point to be shaded, the amount of local visibility (or respectively, occlusion value) over a hemisphere oriented along the point's normal and to apply this value to reduce the incoming light in a reflection equation. Considering such incoming light reduction as a local phenomenon, this visibility is often limited to a given maximum distance. This visibility is sometimes also tapered with a monotonical function of distance

Figure 5.18. Rendering using ambient occlusion. *Image courtesy of NGRAIN (Canada) Corporation.*

within the range $[0,1]$ to simulate how a more distant object blocks less of the incident light. These two controls allow artists to modify the extent of ambient occlusion present in a scene for different scales of scenes. This alternate tapered function to local visibility is distinguished by some by the name obscurance. We will not discriminate between obscurance and ambient occlusion in the following, as both can be applied to each technique with small changes.

The occlusion value can be precomputed or computed on the fly, and can be stored per original vertex (per resampled surface points) in a special occlusion texture mapped over the surface or in image-space from a given viewpoint. In order to obtain more precise shading values, directions associated with the non-occluded portion of the hemisphere above a point can also be added. This requires extra storage for one average visibility direction (known as "bent normal"), or for a number of such directions (stored in one map, a cube-map, or compressed as wavelets or spherical harmonics), all represented as bitmasks or depths. In Figure 5.19(center), the visible portions above a point P_i on the surface are sampled by rays. The

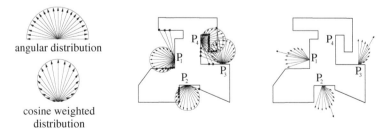

Figure 5.19. Two distributions of rays (left). Rays sampling local visibility at each location P_i (center). Unoccluded rays and resulting bent normal at each location P_i (right).

unoccluded rays contribute to the average visible direction of the local bent normal (Figure 5.19(right)).

Because computing the local visibility is the key to ambient occlusion, several methods have been developed and tuned to this process and to application requirements. They are often divided into object-space and screen-space methods, and their computations use either ray casting or GPU rasterizing. However, several object-space strategies also apply to screen-space strategies and vice versa.

5.8.3   Object-Space Methods

Ray casting methods provide the most general solution and produce the best results for film production images. However they are also the most time-consuming solution as many rays need to be shot (high-quality production may often require up to 1024 rays per point). Since the rays can be limited in distance and they are coherent in origin and distribution (often following a cosine-weighted distribution according to the surface normal), several optimizations can be exploited. This is also the case for computing an ambient-occlusion factor over an entire polygon instead of at a single point. Rays are then distributed in location over the polygon, as well as in angular direction.

For instance, considering that ambient occlusion is a rather low-frequency phenomenon, Reinbothe et al. [477] ray trace voxels more efficiently than the actual scene geometry. Laine and Karras [327] precompute a low discrepancy distribution of directions and use the same distribution for all points to be shaded. At each such point, they slightly perturb (rotate) the point's tangential and normal orientations in order to reduce banding artifacts. Their optimized bounding volume hierarchy (BVH) traversal is inspired by the BVH ray packets of Wald et al. [612] (see Section 2.5.2) and achieves an important speedup.

Using the power of GPU rasterization to sample visibility instead of using ray tracing has also been popular (e.g., [460]).

Bunnell [76] replaces each vertex of the scene polygons with a disk of a radius related to the size of the polygons shared at the vertex. Ambient occlusion is approximated by the form factors (see Section 5.10.1) between all disks. A similar approach to clustered hierarchical radiosity [544] and lightcuts [614] is used to reduce the $O(n^2)$ computations to $O(n \log n)$, where n is the number of radiosity patches, point samples, or disks. In this approach, a pair of disks with larger formfactor values produce more (ambient) occlusions on each other. Because between a pair of disks, Bunnell does not consider occlusion due to other disks, a disk lying behind a closer disk will still contribute to the ambient occlusion of the first disk. Therefore, a first pass between all disks results in a darker ambient occlusion. To reduce the contribution of a hidden disk, Bunnell multiplies in a second pass each form factor of a disk by its own local ambient occlusion. Hence, strongly occluded disks contribute less to the ambient occlusion of other disks. More passes of this computation reduce occlusions of occlusions. Figure 5.20 illustrates this principle.

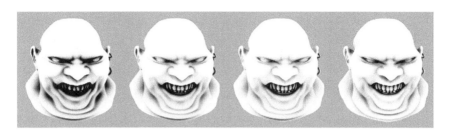

Figure 5.20. Multipasses of form-factor computations can reduce ambient occlusion over-occlusion. From left to right: one pass, two passes, three passes, and 200 random rays per vertex. *Image courtesy of Bunnell [76], and model courtesy of Bay Raitt.*

Hoberock and Jia [246] suggest a number of improvements to reduce the arti-facts (e.g., interpolating ambient occlusions between levels in the hierarchy, shift-ing disks from a vertex to the center of polygons, and computing real form factors for very nearby polygons). Unfortunately, these methods still tend to overshadow the elements, and do not scale well for complex scenes.

Another class of methods encode into a cube-map around each object an occlusion field that would be produced by this object on its surroundings. Malmer et al. [379] precompute a spherical occlusion factor (in the form of a di-rection and a cone of ambient occlusion) at each center of a 3D grid encompassing an object. The grid is extended according to an estimation of the influence of the object on ambient occlusion. For grid points falling inside the object, the full am-bient occlusion is replaced by interpolated values from the surrounding voxels. At rendering time, ambient occlusion due to an object is fetched from the point to be shaded transformed in the object's 3D grid, and the cone is clipped/weighted by the surface normal. Values are typically interpolated from the 3D grid. The values near the borders of the 3D grid are rescaled to reduce artifacts. Kontkanen and Laine [315] propose to compute the occlusion factor due to an object for points distributed around this object, each occlusion factor being stored in the form of a sphere cap and a direction of occlusion. They also compute a correction factor to capture how the direction of occlusion deviates from the center of the object as a point gets closer to the object. Instead of storing these values in 3D space, they sample them along each direction of a cube-map from the center of the object and fit the data to a quadratic equation as a function of the radial distance. For a point to be shaded within the convex hull of the object, an approximation formulation is used. In the two methods, ambient occlusion values from the occlusion field of nearby objects are approximated by blending them together (multiplication) for each point to be shaded. These methods rely on precomputations and may result in large memory requirements, but since they aim at capturing lower-frequency am-bient occlusion, their resolutions are usually set to reasonably small values. They lend themselves well to animation for rigid-body transformations.

Laine and Karras [327] devise a scheme based on a tight bounding volume for each triangle in order to scan-convert it into a cube-map around each point to be shaded to identify which of its low-discrepancy directions are blocked by this triangle. Although less efficient than their ray casting solution in the same paper [327], Laine and Karras' method shows much improvement compared to an efficient ray casting solution.

McGuire [401] analytically computes the occlusion factor at a point to be shaded for each triangle in a scene. He builds some kind of "ambient shadow volume" around each scene triangle to determine efficiently which triangles could contribute to an ambient occlusion factor on a 3D point. While very precise, benefiting from a two-pass GPU rendering, and providing high-quality results, the method can result in some overocclusion when the projection of several triangles overlap on the hemisphere at a point.

As a side note, even though several methods discussed above rely on data stored in a cube-map in order to determine directional information at a given point, spheres and spherical shapes have also been frequently used as approximations of objects for efficient occlusion computations (e.g., [478]). They have been applied as well for ambient occlusion to compute and store occlusion with spherical harmonics [541].

5.8.4   Screen-Space Methods

Screen-space methods (called screen-space ambient occlusion (SSAO)) compute a depth buffer from the viewpoint and use only these values (depth and sometimes surface normal) in order to determine ambient occlusion. Surfaces hidden or undersampled from the viewpoint may be missed, resulting in visibility errors. However, because the depth buffer data is readily available on the GPU after visibility determination, this approximation of the scene (from the viewpoint) can provide a reasonable solution that is fast to compute and independent of scene complexity. This is why several interactive and real-time applications, including the demanding game industry, rely mostly on screen-space ambient occlusion.

The basic methods use the depth buffer from the viewpoint with or without surface normals, generate a number of samples within a sphere (centered at the depth of the pixel, with a radius related to the pixel distance), and test against the depth buffer how many of these samples are visible [415, 279]. Full visibility is achieved when less than half of the samples are occluded, i.e., when they fall below the visible hemisphere, assuming a local flat surface. Increasing the number of samples reduces the resulting noise in the ambient occlusion value, but this has a direct impact on rendering performance.

For a point to be shaded, rays can be traced in 3D and tested at regular intervals against the depth buffer, i.e., marching up to the maximum distance (sphere radius). If a ray gets behind a depth for a depth-buffer pixel, it is considered as intersecting the scene and therefore contributing to the occlusion. While this ap-

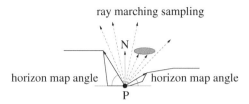

Figure 5.21. Even with a horizon-map structure, other occlusions can be missed. Bavoil and colleagues propose a ray-marching sampling [38, 42].

proach is reliable, it is improved when the data is organized in a highly coherent depth map.

In a method similar to Sloan et al. [541], Shanmugam and Arikan [521] assume that each pixel in the depth buffer is a 3D sphere of radius related to its projected pixel size and centered at its depth. For a given pixel, the ambient-occlusion value corresponds to the projection of all such spheres within a maximum distance. The method may suffer from overocclusion since overlapping projections are not treated.

Instead of point samples in a sphere, Loos and Sloan [364] use sampling lines (and areas) and a statistical model of the local depth in order to better estimate the visibility. By improving the position of samples over these lines, they improve the quality of the reconstructed visibility with less samples.

In horizon-split ambient occlusion, Bavoil et al. [38, 42] assume the depth buffer is a heightfield and compute a horizon map (e.g., [391]) at each pixel location. For each tangent direction (slice), the horizon angle is determined by traversing the corresponding adjacent pixels in the depth buffer and updating the corresponding horizon angle. The ambient occlusion due to the horizon map is approximated by connecting the adjacent horizon angles and integrating the visible angle over the hemisphere. However, the horizon for the normal direction could miss other occlusions within the maximum distance, as shown in Figure 5.21. Ray marching is used between the horizon angle and the normal direction, distributing the angles according to the cosine factor. It is also possible to compute the horizon for an entire sphere, neglecting the normal direction. The contributions to occlusion from below the hemisphere at the normal are simply canceled out.

Some of the data missed with a single-layer depth map can be reduced with multiple depths per pixel, implemented as depth peeling [39, 482]. For a given scene with a high depth complexity from the viewpoint, the resulting computations can, however, be significantly slow. Polygons at grazing angles with the view direction can also introduce wrong results. Another solution to reduce the missing data consists of using multiple viewpoints [482]. Each position and view direction must, however, be set with care in order to provide useful data without adding too much rendering.

Another problem with missing data in SSAO occurs at the borders of depth buffers because of the missing depth values outside the view frustum. Bavoil and Sainz [39] propose to extend the view frustum by constant size or by an estimate of the maximum depth for a ray according to the current depth at the border pixels.

In order to increase performance, Bavoil and Sainz [39] suggest using lower resolution for depth and normal data, thus improving cache efficiency. Upsampling with edge-preserving filters then fills in some of the data. They also suggest doing two passes, one at half resolution, from which the local variance is checked, and if the variance is higher than a threshold, ambient occlusion is computed at full resolution.

Hoang and Low [245] propose capturing closer and farther ambient occlusion values by sampling only with one small kernel but in multiresolution depth buffers, and then combining the computed ambient-occlusion values. Bilateral filtering and upsampling at coarser resolutions help to reduce noise and blur artifacts, and the small kernel offers efficient GPU fetches.

Loos and Sloan [364] propose considering that every depth in a pixel has a given thickness. This improves the quality of the ambient occlusion effects because silhouette edges do not contribute anymore to more distant adjacent background surfaces.

In volume ambient occlusion, Szirmay-Kalos et al. [574] reformulate the directional hemispherical integration of ambient occlusion into a volume integration of a (smaller) tangential sphere. They must assume that, at most, one intersection can occur for one sample in the associated tangent sphere, and they lose the ability to capture both distant and close ambient occlusion phenomena. This remains as efficient as the best SSAO methods while producing more accurate ambient occlusion with fewer samples. Interleave sampling is used to reduce noise or banding effects due to the sampling of the new formulation.

Instead of computing an ambient occlusion factor that multiplies the incident illumination, Ritschel et al. [482] take each visible point sample generated in a hemisphere in a basic SSAO approach and use the corresponding direction from the point to be shaded to identify the incident illumination, thus computing the associated shading value. The occluded samples contribute to darken their section of the hemisphere. The authors also project each sample to its closest surface in the depth buffer and use these points to compute a local one-bounce indirect illumination.

5.8.5   Animation

In general, the naive implementation of ambient occlusion is only suitable for static walkthroughs. Akenine-Möller et al. [7] discuss techniques to allow fast computation of dynamic changes over an animation.

The efficiency of screen-space ambient-occlusion methods position them well for animated scenes since the solution is recomputed at every frame. Unfortunately, these methods suffer from temporal discontinuities due to undersampling problems or from performance issues when more data (higher resolution, multiple depths, etc.) is generated.

In most animated scenes, only a few objects move while most of the scene remain static. It is therefore only necessary to recompute the ambient occlusion affected by the movements [404].

Methods based on a form of occlusion field (e.g., [315, 379]) are also appropriate for dynamic scenes where the objects are affected only by rigid transformations.

When deformations occur in a dynamic scene, Kontkanen and Aila [314] consider the specificities of an animated character, compute ambient occlusion values for some poses, and identify correspondences between joint angles at vertices. They then interpolate linearly according to the correspondence matrix. Kirk and Arikan [308] compute similar correspondences but also compress the large data thus obtained.

While interesting for each individual character, these solutions do not easily extend to computing ambient occlusion between characters nor different types of objects.

5.8.6 Trends and Analysis

In only a few years since their introduction, ambient-occlusion techniques have become integrated in most of the image production applications, as much in high quality renderings such as in film productions as in real-time renderings for video games. Most commercial rendering software as well as game engines offer an implementation of ambient occlusion. Even when the results of some ambient-occlusion techniques are only approximations with associated known limitations, the increase of realism that they provide is considered beneficial in most applications.

Very recent improvements on multiresolution sampling have helped much in reducing several artifacts. It is expected that the accelerated pace of new solutions will go on for at least the next few years.

5.9 Precomputed Radiance Transfer

Precomputed radiance transfer (PRT) approaches compute and store light-transport data, such as point-to-point visibility and reflectance, in a compact form using various basis-space representations. After this costly precomputation, the light-transport data can be used to very efficiently relight the scene with a new external lighting configuration. As such, PRT can be viewed as a trade-off between fully offline and fully dynamic real-time rendering solutions. The more costly the precomputation and storage in PRT, the more realistic the results, but the slower the performance. The key idea behind PRT is that in fixing one or more properties of the incident directional illumination, reflection model, interreflections, occlusion, or geometry, and representing all these properties in a common basis-space, it becomes possible to rerender a scene from the cached information using simple GPU-accelerated operations and do this with real-time performance and high quality.

Examples of such data include infinitely distant incident illumination (e.g., from an environment map storing the incident light radiance for all directions), isotropic reflection functions (BRDF), and binary mask to encode point-to-point visibility. Uncompressed, this data quickly becomes unmanageable to store and manipulate (especially in the context of real-time rendering systems). PRT takes advantage of (potentially hierarchical) basis representations to compress this data. Furthermore, the choice of which basis to use incurs its own performance versus accuracy trade-offs. Spherical harmonics (SH) are a popular choice for PRT since they very efficiently encode smooth, low-frequency signals (e.g., smooth skylight, diffuse reflectance, smooth soft shadows) with only a few SH coefficients. Haar wavelets [423], spherical wavelets, and spherical radial basis functions (SRBFs) [593] are basis functions that are more adapted to higher frequency signals that can handle more accurate so-called *all-frequency* shading effects, such as the combination of both smooth and hard shadows. However, while better suited when compared to SH for these higher-frequency effects, these basis representations require more storage and can require more work to process at runtime.

At each (discrete) point of a scene, all incident lighting must be integrated by the rest of the scene. By choosing a basis with appropriate mathematical properties to express light transport, these integrals can be very efficiently computed using simple dot-products, or matrix multiplications, of compact coefficient vectors.

Sloan et al. [539, 295] introduce PRT and establish many of the fundamental concepts behind PRT approaches. By assuming both low-frequency environment lighting and reflection (BRDF), light transport signals can be encoded with only a few SH coefficients, and shading can be computed in real time for (static) scenes consisting of thousands of vertices. Transfer functions at each vertex can include visibility for self-shadowing and interreflections; however, only low-frequency results are possible with this representation. They also show results of objects casting soft shadows over terrains, and low-frequency caustics from specular interreflection.

Several authors have addressed the low-frequency restriction, among which, Ng et al. [423] propose the use of Haar wavelets to encode the light transport signals. Their method handles all-frequency signals much more efficiently than SH, requiring typically two-orders-of-magnitude fewer coefficients to handle higher-frequency shading effects, thanks to the local support of the wavelet basis. In particular, shadows can be finely determined, assuming an as-fine mesh. Despite these advances, PRT techniques still require a large amount of data to be stored over a scene, and a number of other schemes (e.g., [540]) have attempted to further com-

Figure 5.22. Comparing uniform (left) and adaptive subdivision (right) of the mesh for precomputed radiance transfer. ©*2004 ACM, Inc. Included here by permission [320].*

press these representations. We refer the interested reader to the survey of Ramamoorthi [474] for more details.

Regarding shadowing, as with radiosity approaches (Section 5.10.1), a fine level of tessellation is required for PRT approaches to properly capture shadow boundaries. Alternatively, PRT data can be stored in texture space, in which case the texture resolution must be suitably high to capture shadow boundaries. Unfortunately, discontinuity meshing techniques cannot be easily applied to PRT. Křivánek et al. [320] propose an adaptive method to subdivide a mesh by detecting large changes in the transfer operators at the vertices in order to reduce the artifacts at shadow boundaries. See Figure 5.22 for the improvement of adaptive subdivision over uniform subdivision of the mesh.

One of the major limitations of standard PRT techniques is that they are restricted to static scene geometry since the precomputation that takes place in the scene is not computable in real time. To loosen this constraint, Zhou et al. [673] introduce "shadow fields" around rigid objects, storing PRT data that encode how the object contributes to the occlusion of its surrounding space. At runtime, the shadowing at a given point is computed by combining the occlusion fields from each (dynamically animated) rigid object, after which the standard PRT algorithm can be used to reconstruct the shading at the point. The spatial and angular sampling rate of each object's shadow field varies depending on whether low- or all-frequency shadows are required, as well as on the maximal distance within which the object can cast a shadow. This latter issue can be exploited at runtime to ignore objects far away from points to be shaded. Finally, the occlusion data in the shadow field is compressed with SH or wavelets.

One major advantage of this representation is that a shadow field can be shared by all the instances of an object, and the idea easily extends to local light sources. Unfortunately, shadow fields suffer even more seriously from storage limitations, requiring hundreds of megabytes for even simple blockers. Moreover, combining basis-space occlusion for all blockers (at each point to be shaded) can become quite costly. Ren et al. [478] address these two limitations of shadow fields, and offer support for non-rigid blockers by representing blocker geometry with a hierarchical sphere representation. They compute and accumulate basis-space occlusion across the blockers in a logarithmic space and efficiently exponentiate the basis-space occlusion before applying standard PRT reconstruction. Their approach is much more efficient and requires very little storage. Their results are slightly more approximate than standard low-frequency SH PRT since the logarithmic and exponential operators introduce errors into the resulting SH reconstruction of visibility.

5.9.1   Trends and Analysis

Precomputed radiance transfer capabilities have been included in DirectX since Version 9, exposed through the ID3DXPRTEngine APIs. Various forms of PRT have appeared in video games such as Halo 2 and Half-Life. However, adoption has been slow since limitations to rigid scenes, tessellation/sampling requirements, and storage costs do not yet align with the use cases in current video game design workflows.

5.10   Global Illumination

In this section, we continue with soft shadow coverage, but with our focus on soft shadows due to global illumination. We cannot possibly cover the entire body of global illumination literature, so we focus on the overall approaches and elements of the literature that have impact on shadows in particular.

Global illumination algorithms take into account both the direct illumination from a light source (discussed in the previous chapters) as well as the indirect illumination of light that interreflects/interrefracts throughout an environment, i.e., light bounces off a wall multiple times before reaching the camera. The mathematical foundation of this is described in the rendering equation [282]. It is the indirect illumination that softens the shadow boundaries. Figure 5.25(left) shows the direct illumination and Figure 5.25(right) shows the shading of a point on the floor with secondary illumination; the shadowing is soft because there are now implicitly multiple indirect sources. Global illumination algorithms tend to provide more realistic solutions, but are computationally more expensive than direct illumination algorithms (which are most of the algorithms discussed in this book so far).

The global illumination discussion is divided up into two categories: radiosity techniques, followed by Monte Carlo ray tracing techniques. Radiosity is well suited for purely diffuse environments and requires deep tessellation into mesh elements. Because there tends to be no specular component in radiosity-based algorithms, the illumination can be precomputed and its results used without alteration during shading in real-time walkthrough applications. On the other hand, Monte Carlo ray tracing algorithms are more flexible, require no deep tessellation, and are good for diffuse and specular environments, and appropriate for more than just static walkthrough scenarios, but they are generally not real-time appropriate, and thus tend to require offline renderings, especially in film and static-image situations.

It is sometimes difficult to distinguish a direct illuminated rendering of soft shadows from extended lights from a scene rendered with global illumination. This can provide some heuristics that allow good placement of a few point lights or extended lights to generate the global illumination look [174]. These heuristics are taken advantage of in instant radiosity and virtual point-light approaches, which is why a third category has been added to supplement the radiosity and Monte Carlo approaches.

5.10.1   Radiosity

Common among many radiosity algorithms, the surfaces of the scene are tessellated into many smaller patches, where each patch theoretically represents a small region where a unique radiosity value is computed. A form factor is computed for each pair of patches. Patches that are far away from each other, oriented at oblique angles relative to one another, or are mutually occluded, will have smaller form factors. The form factors are used as coefficients in a linear system of equations. Solving this system yields the radiosity, or brightness, of each patch, taking into account diffuse interreflections and soft shadows, but solving this system tends to be slow [200], thus progressive refinement approaches tend to be the norm [104], and patch-to-patch visibility calculations then result in the eventual shadowing result.

In a walkthrough animation scenario, assuming only diffuse interreflections, solving this system can be done just once as a preprocess, and the only shading computation required during actual rendering is the Gouraud shading of the patch information. This has been the most common application of radiosity approaches, and it is appropriate for architectural and interior design applications.

The reason for the smaller patches is that a single patch indicates a single value for illumination, which means that if the patch is not small enough to capture high-frequency changes (usually due to shadows), then visible blockiness artifacts will occur, which is why *discontinuity meshing* has been an important research area in radiosity algorithms: it represents the optimal determination of patch sizes placed at the right locations to capture high-frequency changes. By identifying patch boundaries along the shadow discontinuities in the zeroth (zeroth derivative only exist for light sources that can produce a hard shadow discontinuity, such as a point light or a linear source that is parallel to the edge of an occluding object), first, and second derivatives and intersecting them with the scene, radiosity solutions can converge faster at a higher visual quality. This is because the patches will not

Figure 5.23. Improvement of radiosity image quality after applying discontinuity meshing. ©*1992 IEEE. Reprinted, with permission, from [353].*

exhibit blockiness artifacts due to Gouraud shading, as the shadow discontinuities are set at the patch boundaries. This also avoids shadow leakage where a shadow extends on mesh elements straddling illuminated and shadowed regions. Important work in this area comes from the following papers [80, 234, 353, 19]. See Figure 5.23(left) and (right) for a before and after set of images respectively, that show the consequence of brute-forced subdivision into patches and the application of discontinuity meshing.

To compute even more accurate irradiance on top of discontinuity meshing techniques, back-projection techniques are used to build a complete discontinuity mesh [560, 141, 561, 142, 368, 101, 32]. The back-projection in a region contains the set of ordered lists of emitter vertices and edge pairs such that for every point to be shaded, the projection of those elements through this point onto the plane

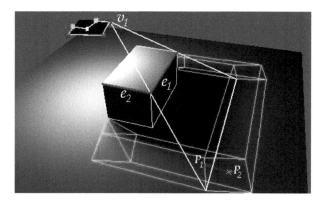

Figure 5.24. Back-projection as applied to discontinuity meshing. ©*1994 ACM, Inc. Included here by permission [141].*

of the light source forms the back-projection instance at this point. The partition of the scene into regions having the same back-projection is what forms the complete discontinuity mesh (see Figure 5.24). Given the complete discontinuity mesh, the penumbra irradiance can be efficiently computed. The different papers (listed above) utilize back-projection and propose optimizations on top of the back-projection technique.

There are alternative algorithms that produce high-quality shadows within radiosity algorithms without resorting to some form of discontinuity meshing:

- Zatz [656] uses 40 × 40 shadow maps with sampled and interpolated visibility values as a shadow mask to come to a high-order Galerkin radiosity solution. However, the determination of a good shadow map resolution and placement of the shadow maps are not discussed.

- Soler and Sillion [551] use texture convolution [550] to produce a visibility texture applied to the mesh as a shadow mask.

- Duguet and Drettakis [144] use a robust method to compute shadow boundaries analytically without a need to generate a subdivided mesh, and small shadow features can be merged while preserving connectivity.

5.10.2   Virtual Point Lights

Instant radiosity techniques [299] strategically place a number of virtual point lights (VPLs) to simulate the soft shadowing from the emitting light sources and from the indirect illumination. The shadowing is achieved using either shadow depth maps [116, 329, 481, 247] or ray tracing [516, 517], with the optimal placement of the virtual point lights and visibility tests as the common challenges in avoiding light leaks. Such a simulation of virtual point lights can be seen in Figure 5.25, where the left image shows direct illumination and the right image shows secondary illumination based on virtual point lights.

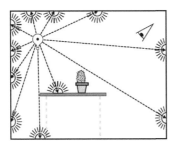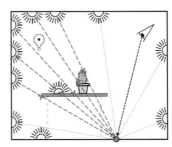

Figure 5.25. Instant radiosity with placement of virtual point lights to achieve shadowing. *Image courtesy of Laine et al. [329], ©Eurographics Association 2007. Reproduced by permission of the Eurographics Association.*

Walter et al. [614] construct a hierarchical representation for VPLs (emitted and reemitted), where one VPL (or a few, to allow for stochastic selection) gets the intensity of all its children VPLs. At the point to be shaded, they test in a top-down fashion (a cut in the tree of VPLs) when the combined VPL contribution is large enough to deserve sending shadow rays. Bounds on each shading term allow efficient evaluations of both local VPLs and global, more intense VPLs. The method is conservative, but because tight bounds on shadowing are difficult to efficiently compute, all VPLs are simply assumed to be visible. This can lead to oversampling of intense VPLs that are occluded from the point to be shaded.

Hašan et al. [229] express this many-light problem in a large matrix of image pixels in rows and VPLs in columns. Rows and columns are reorganized into similar clusters to efficiently subsample the VPLs. The method is appropriate for intense VPLs, but local VPLs are sampled for all pixels. Ou et al. [443] improve on the clustering in order to achieve a good balance between local and global VPLs. While shadows were computed via shadow depth maps in their original work [229], because of bias and sampling artifacts, they rely on shadow rays for their later work [443].

5.10.3   Monte Carlo Ray Tracing

Monte Carlo techniques are often applied to ray tracing, and in the case of global illumination, it can mean one of several things:

- Many chosen rays are shot from the light and interreflected onto the scene, and this contribution is accounted for in the rendering when it hits the camera. This is sometimes referred to as backward ray tracing [18, 86]. To hold indirect illumination information associated with points on a surface, textures per surface [18] or a voxel structure [86] can be used. Shirley [529] uses a three-pass solution, the first pass using backward ray tracing to capture specular interreflections, the second using radiosity to capture diffuse interreflections, and the third using standard ray tracing to calculate direct illumination. This mixture results in convincing images such as the one seen in Figure 5.26, including caustics effects.

- Rays are traced from the camera to the scene and followed until they reach a light source or connect to a light source. This is often referred to as path tracing [282]. Acceleration of this approach can be done through irradiance caching [624], where the indirect illumination can be derived by interpolating cached irradiance values.

- Variants of the first two techniques connect both types of ray paths (from the light and from the camera) with careful weighing according to the probabilities that such paths exist. This is often referred to as bidirectional path tracing [324], with Metropolis light transport [602] being a common usage of bidirectional path tracing.

Figure 5.26.   An early image illustrating the rendering of caustics.   *Image courtesy of Shirley [529].*

- Photon mapping [270, 272, 273, 384] is a special case in which rays from the light source are shot and interreflected until a diffuse surface is hit. The photon results are stored in a photon map. The final render does a single camera view pass and reconstructs the irradiance with the closest photons.

In all the above approaches, shadows are implicitly considered due to the tracing of the rays with first visible hits in mind. In other words, shadows on a surface exist due to reduced amount of visible hits from the interreflections.

5.10.4   Trends and Analysis

Because there tends to be just the diffuse component in radiosity-based algorithms, the (view-independent) illumination can be precomputed (or "baked"), making it very useful and often seen in architectural and interior walkthrough applications. Example software that provides radiosity capabilities include Lightwave, Lightscape, Renderzone plus, Englighten. Instant radiosity and virtual point-light approaches seem to be the most researched approaches in the recent work.

On the other hand, Monte Carlo ray tracing algorithms are flexible, tend to require offline renderings, and have become useful in film and its rendered animations seen in CAD, automotive, and architectural visualization. Example software that includes Monte Carlo ray tracing includes Mental Ray, Maya, LuxRender, etc. Not one single approach (among the four categories listed above) has been singled out in terms of adoption, but often, several approaches are implemented in a single system for the user to choose.

5.11   Final Words

This chapter has covered other treatments of shadows, most of which can be considered as advanced features of the basic shadow algorithms. The importance of those advanced features is significant, such as atmospheric shadows and motion blur in film and video. Also, the investment of R&D into ambient occlusion, precomputed radiance transfer, and global illumination has overtaken the interest of the soft shadow "look" due to extended lights.

In the next chapter, we cover some applications of the shadow algorithms.

CHAPTER 6

Applications of Shadow Algorithms

6.1 What's Covered in This Chapter

There are many standard uses of the shadow algorithms described in this book, with the main aim of generating photorealistic images of digital 3D models. In this chapter, we discuss other applications of shadow algorithms, such as

- Integrating 2D images with 3D graphics, where 2D represents live action, and 3D represents digital creation. This is often referred to as augmented reality (Section 6.2) [433].

- Non-photorealistic environments to simulate sketch-like or paint-like, artistic renderings (Section 6.3) [383, 501, 278, 118].

- Shadows as interaction tools, where the computation of shadows is used not as a rendered result, but as a tool to guide other aspects of interactions in a 3D scene (Section 6.4).

6.2 Augmented Reality

There has been much research done in the area of acquiring information from static or live-action 2D images (using computer vision techniques), in order to incorporate new 3D graphics geometry to be rendered together so the integrated elements will all appear photorealistic and indistinguishable from the origins of their representations. This is often referred to as "augmented reality" in the literature and has become an important application in film, medicine, manufacturing, visualization, path planning [31], etc.

Historically, some augmented-reality aspects were integrated into DCC software (offline purposes) as early as the 1990s. However, there was little recorded computer graphics literature on the subject until Fournier et al. [182]. Use of real-time techniques were not published until 2003.

The major chores in integrating new 3D geometry in 2D images include

1. Deriving the view configuration from the original image so that new geometry can be added accurately into the scene, phantom objects (see item 3) can be accurately reconstructed based on the original image and additional input for the illumination aspect of the rendering equation [282]. For a video sequence, it is necessary to track the motion of the camera and possibly of some objects.

2. Deriving some lighting information (some from analysis of shadows) [497, 305, 498, 288, 120, 378] in order to support the proper rendering look of the new geometry and to provide light directions for shadows, which has been considered critical to the success of final integrated images [568, 377]. Otherwise, the new geometry will look unrealistic and out of place. An example can be seen in Figure 6.1, where the torus is the new synthetic geometry inserted in the rest of the scene captured by the original image; without the shadows from the torus on the can and table, the torus stands out as something that does not quite fit in the scene.

3. Reconstructing phantom geometry, where phantom geometry represents what should be the geometry present in the original image (but not represented in 3D). This reconstruction serves two purposes. The first is to provide relative depth comparison to the new geometry, i.e., the phantom geometry is either in front of or behind the new geometry. The second is to provide geometry reference for proper shadowing. Note that this reconstruction usually requires manual input (often using image tracing combined with photogrammetry or other techniques (see Section 6.2.3)). The objective is to minimize the number of phantom geometries that need reconstructing, i.e., if it represents a background object and has no implications on the shadowing from and of the new geometry, then the reconstruction should be avoided. Also, if phantom geometry is needed to partially occlude the new geometry from a camera viewpoint, then the reconstruction needs to be detailed, whereas if phantom geometry only has an impact on shadowing, then the reconstruction can be approximate. Note that this reconstruction is usually the most time-consuming step to achieving augmented reality.

6.2.1   Augmented Reality Assuming Direct Illumination

From a shadowing standpoint, phantom geometry acts as an invisible object for receiving and casting shadows. It only paints itself darker when there is a shadow cast on it by the new geometry. Also, it should ideally only cast shadows on other

Figure 6.1. Augmented reality scene without and with shadows accounted for by the synthetic object (torus) and the phantom objects (can, table). ©*2003 ACM, Inc. Included here by permission [222].*

new geometry because shadows from phantom objects onto other phantom objects are already accounted for in the original images. However, this distinction makes the shadow algorithm difficult to implement if a ray tracing approach is not used—checking is needed per point to be shaded whether the source of the shadow is from a phantom object or not. Therefore, a simplification for this phantom geometry is to only receive shadows (and assume it will not cast shadows on the new geometry); thus, the alternate name of "shadow catchers," "matte objects," or "holdout objects" in certain applications such as Sketch!, Cinema 4D, ProAnimator, and Blender. Note that in the example of Figure 6.1, the phantom objects only receive shadows.

Avoiding Double-Counting Shadows

As hinted above, if phantom objects do cast shadows on other phantom objects, such shadows would be double-counted and may not visually match what is already there. Another form of double-counting exists even if the above case is properly handled, in which the phantom object is already shadowed in the original image, but the new geometry also occludes it. For this case, there needs to be some registration not to occlude the already occluded. An example of double-counting can be seen in Figure 6.2, where a single light seems to cast multiple intensity shadows.

To resolve the two double-counting shadowing problems, Jacobs et al. [262] use

Figure 6.2. Double-counting shadows from a single light source. Image courtesy of Jacobs et al. [262].

computer vision techniques to detect shadows in the original image and then protect that region from interference of newly rendered shadows. Another variant is to remove the shadows from the original image and then to render (receive and cast) shadows for all new geometries as well as phantom objects [378, 284, 371]. However, with increased complexity of the scene, the accurate detection and removal of shadows can be in itself a very difficult task.

Hard Shadows

A number of authors use shadow depth maps [559, 568] to achieve augmented reality, while others employ a variation of stencil shadow volumes [222, 262]. Shadow volumes are appropriate because the reconstructed phantom objects should be approximate and few; thus, geometry complexity should not be high. However, only hard shadows are generated and can be a giveaway in terms of mismatch of the original image versus graphics rendering. Figure 6.1 shows how the inclusion of shadows helps to register the synthetic objects in the scene. The darkening due to the computed shadows onto different surfaces does not match the original image's softer shadows.

Filtered shadow depth maps using percentage-closer filtering [476] can be effectively used. However, care must be taken as to how boundary cases for a spotlight are handled. When a filter region for shadow depth map comparison is done, if the shadow map pixel is outside the light frustum, the computation should return it as in light, although non-augmented-reality renderings should indicate that it is in shadow. This is because the lighting and shadowing outside the light frustum should already be accounted for in the original image, or a dark rim around the boundary of the light source will result.

Soft Shadows

Kakuta et al. [283] use the concept of shadow catchers for phantom objects, and render and store a hemisphere of lights and shadows so that soft shadows can be achieved in real time, where soft shadows are a result of the linear combination of a few of those lights and shadows.

Image-based lighting techniques [120] have used divisions of an environment map over the scene, with each division containing a light source that can simulate incident lighting. Building on top of this technique, several authors [571, 376, 204, 422] use multiple shadow depth maps over this environment map to render soft shadows. Similarly, Noh and Sunar [434] use the soft shadow texture technique [240, 232] (Section 4.4.1) to blend hard and soft shadows together effectively.

6.2.2   Augmented Reality for Global Illumination

Several authors [182, 143, 369, 194, 195] use an interactive, cluster-based radiosity system to generate shadows cast by the new geometry. In a radiosity-based global illumination environment, this means that every phantom object in the original

image needs to be reconstructed in 3D (often with some form of manual intervention) because the additional 3D geometry will affect the illumination of everything
in the original image's objects. As with the direct illumination case, the phantom
object can paint itself darker when shadows are cast on it, but unlike the direct illumination case, the phantom object can also paint itself differently with additional
illumination from the new geometry (via interreflections).

Debevec [119] splits the scene into local and distant sectors. Distant sectors can
be unaffected, and thus no reconstruction is needed. Only phantom objects in local
sectors need reconstruction. Ray tracing and differential rendering techniques are
used to produce the integrated results.

6.2.3   Model Reconstruction Techniques

The most common techniques suggested in the literature for geometry reconstruction use some variation of photogrammetry—available software using such techniques include RealViz, ImageModeler, Photomodeler, PhotoScan, etc. Some additional reconstruction techniques are listed below.

Petrović et al. [458] generate approximate shadows from hand-drawn line art
by approximating 3D polygonal meshes from hand-drawn animation acting as
shadow casters. Tools are available for the user to manipulate the meshes, while
making sure that the image-space silhouettes still match the drawings.   Input
is needed from the user pertaining to the relative 3D positions of the drawing's
shadow casters and receivers. The shading and shadows of the 3D geometry are
composited with the drawings to form realistic shadows from hand-drawn animations. The shadows are generated using ray tracing; however, they can be generated
using any of the algorithms listed in this book.

Chuang et al. [102] attempt to solve a slightly different problem in integrating
the original images with 3D graphics geometry. The goal is to extract the shadow
matte produced by some main object, assuming the shadow is cast on a planar
ground. This shadow matte goes through a displacement map in order to con-
form to the shape of a new background containing the main object. This displaced
shadow is then composited with the new background, to achieve realistic looking
shadows of the same main object into another background.

6.2.4   Trends and Analysis

There is a superset of augmented reality which includes integration of real/live objects called mixed reality. From a shadow-generation standpoint, the issues in both
augmented and mixed reality appear to be very similar, so we do not cover shadows
for mixed reality specifically.

ARToolKit seems to be a popular software library in which augmented-reality
applications are built, including shadow handling. Although there has been a lot of
literature on shadows within augmented reality, much augmented-reality software

does not support shadows. We suspect the predictability of the automation of accurate shadow generation continues to lack robustness. Within offline rendering production, robustness is not much of an issue because a very restrictive environment is assumed; e.g., the basic "shadow catcher" concept can be used without needing to handle the double-counting issue or any attempt to perform relighting. Example software that contains "shadow catcher" capabilities include Sketch!, Cinema 4D, ProAnimator, and Blender. For a more flexible environment, some of the algorithms described above may be used as a first guess, but there is a lot of manual labor in touching up the scenes afterward due to a lack of robustness.

There have also been a number of applications on intelligent phones that use augmented-reality technology, which are quite impressive but easy to achieve because the view configuration has been determined due to the phone's camera. However, shadows are not considered, likely due to the speed of computation as well as the above complexities of the shadows.

6.3   Non-photorealistic Environments

The term *non-photorealistic rendering* (NPR) is actually a generic term for a few different types of rendering; among them are pen-and-ink (Section 6.3.1), technical illustration (Section 6.3.2), and cartoon rendering (Section 6.3.3). In this section, we go over some usages of shadows in NPR environments, even if there is not a lot in terms of shadow generation to speak of. In fact, detailed shadows can sometimes detract the viewer from the specific areas of focus [501], which is why, sometimes, modified, simplified, or no shadows are preferred in NPR applications.

6.3.1   Pen and Ink

Pen-and-ink techniques usually refer to either line-drawing algorithms, stippling, hatching, or engraving.

Sousa and Buchanan [553] provide a tool for pencil drawing, where a simulation-based approach is used to mimic the characteristics of graphite pencils. Shadows are handled in the same fashion. Tolba et al. [587] generate shadows in a projective 2D representation system. A shadow projection is cast onto a plane, and the user is asked to warp this shadow into the desired location.

Winkenbach et al. [635] and Markosian et al. [382] produce shadows from pen-and-ink renderings by employing something similar to shadow depth maps (see Figure 6.3). A shadow

Figure 6.3. Pen-and-ink rendering with shadows. ©1996 ACM, Inc. Included here by permission [635].

planar map is built in preprocessing, and during display, after the view's planar map has been processed, the remaining visible shadow strokes are clipped against the shadow planar map for the regions in shadow. Martinec [388] uses shadow depth maps, and if in shadow, he uses the same hatching algorithm to render the shadows, but tilted at another angle than from the non-shadowed areas.

More typical shadow displays adopted for pen and ink include shadow volumes binary space partitioning (SVBSP) [634] (Section 2.4.4) and stencil shadow volumes [244] (Section 2.4.2). Shadow volumes are useful for many NPR environments because the simplified geometry allows viewers to focus on the right areas; thus geometry complexity (the usual concern for shadow volumes) is typically not large.

6.3.2   Technical Illustration

There is very little literature discussing shadows for technical illustrations. This is because the objective of technical illustration is to focus on specific areas of importance without being distracted by visual enhancements such as shading and shadows. The only two papers that discuss shadows for technical illustrations only do so for shadows cast on a planar floor.

Gooch et al. [199] simplify the look of shadows by using projective shadow techniques (Section 2.2) on planar floors for non-photorealistic hard shadows. Similarly, for non-photorealistic soft shadows, they use multiple points on the extended light source projected onto different elevations of the floor to compute and then combine the shadows.

Ritter et al. [484] integrate 3D and 2D information displays. The shadows are projected on the floor, and 2D text and annotations occur next to the projected

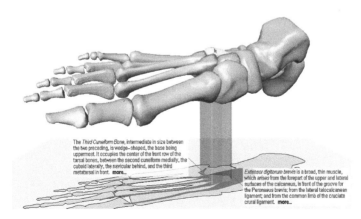

Figure 6.4. Projected shadows for information display. ©2003 ACM, Inc. Included here by permission [484].

shadows to clearly illustrate which sections of the structure are being looked at even though the non-shadow view may have partially occluded the actual structure. Ritter et al. [483] then draw semitransparent shadow volumes of the structure of interest. In this way, the structure of interest is nonambiguous, although it may be partially occluded from view. An example is illustrated in Figure 6.4.

6.3.3   Cartoon Rendering

Lake et al. [331] add color to self-shadowing with a simple self-shadowing function check of $\hat{N} \cdot \hat{L} < 0.5$. This simple self-shadowing can look quite appealing (see Figure 6.5). Spindler et al. [554] introduce stylistic shadows, where the shadows that fall on a foreground object are painted in black, and shadows that fall on a background object are painted in white. This is achieved by adding a foreground attribute in a buffer associated with a stencil shadow volume. Another style is also achieved by drawing a thin outline around the silhouettes of geometry and shadow boundaries.

Figure 6.5. Cartoon self-shadowing. ©2000 ACM, Inc. Included here by permission [331].

Woo [641] introduces an optimization that emphasizes the shadows when the shadow hit distance is small, and causes shadows to fade as the shadow hit dis-

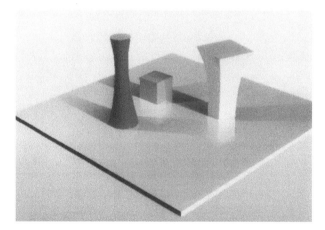

Figure 6.6. Cartoon reflections and shadows. Note the decay in reflections and shadows. *Image courtesy of Woo [641].*

tance grows larger to the extent of entirely culling shadows after a maximum distance maxDist. A simple fading factor can be $((\text{maxDist} - \text{tHit})/\text{maxDist})^k$, where tHit < maxDist. This approach is motivated by cel animation films such as "the alligator and hippo dance" in Disney's *Fantasia*, which is why Woo refers to that optimization as "cartoon shadows" (see Figure 6.6).

DeCoro et al. [125] apply distance transformation and Gaussian blur techniques to achieve (intentionally geometrically incorrect) shadows that are either inflated, with a high level of abstraction, or with softness. See Figure 6.7 for such examples.

Petrović et al. [458] support drawings of shadows in an existing, hand-drawn scene. The user has to specify the depth of various objects in the scene. The approach inflates a 3D drawing of an object based on the hand-drawn art. A standard shadow map is then applied to generate regular shadows. Other typical shadow displays adopted for cartoon rendering include shadow volumes [72], billboard shadow volumes (to render cartoon smoke clouds) [395], and shadow depth maps [122].

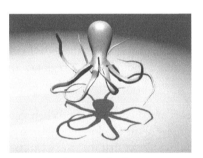

Accurate shadow

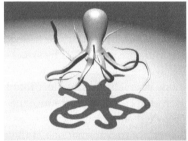

Moderate abstraction
$\alpha = 10, i = 10$

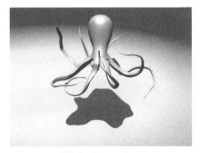

High abstraction
$\alpha = 70, i = 10$

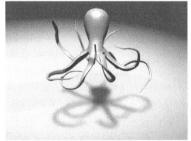

Abstraction and softness
$\alpha = 20, s = 20$

Figure 6.7. Different modes of abstract shadows.©*2007 ACM, Inc. Included here by permission [125].*

6.3.4   Trends and Analysis

This is one of those sections where shadows do not play as important a part. In fact, in certain cases, especially in technical illustrations, shadows can be a distraction in parts of a scene. The main interest in shadow generation appears to be in cartoon rendering, and the shadow techniques described are quite simple.

Additional papers discussing NPR shadows are available, but they are written in Korean [93, 311, 519].

6.4   Shadows as Interaction Tools

Shadows are very useful not only as a graphics rendering effect, but as a tool towards other goals, just as early men used shadows from the sundial to estimate the time of day. In this section, techniques using shadows to position the light or objects in the desired locations are discussed.

Poulin and Fournier [467] suggest using a variation of the shadow volume approach to allow a user to quickly manipulate the geometry of shadows. By modifying the shadow volume while remaining geometrically correct, inverse lighting is computed to indirectly position and aim the light source. Poulin et al. [468] extend their work by instead sketching regions that must be in shadow. Extended light sources are positioned by sketches of umbra and penumbra regions. The resulting system allows the user to quickly position light sources and to refine their positions interactively and more intuitively.

Herndon et al. [241] introduce the concept of a shadow widget. When shadows are projected onto planar walls, transformations of the shadow along some planar walls produce transformations of the occluding object that the user expects. This is done instead of transforming the light, as in the above approaches. Shadows also do not need to be seen as just a darker region, but can use different rendering styles such as silhouette or wireframe modes.

Pellacini et al. [455] describe an interface that allows the user to select a shadow, then move, scale, or rotate the shadow. The system will then compute the necessary changes in either the light or occluding object. Shadow depth maps are used as in the shadow generation algorithm. Pellacini et al. [456] allow the user to paint color, light shape, shadows, highlights, and reflections, as input to compute the lighting and shadowing properties.

Barzel [36] indicates that it can be very difficult to position the same lights and shadows exactly where we want for the desired visual effect. While the above approaches can be used to approximately position lights (or objects), Barzel notes that the shadow direction can be somewhat shifted away from the lighting direction without anyone noticing. Enhancing the lighting model to include a shadow direction to support this can prove to be a big timesaver for the user, especially in a cinematic design environment. Also, opaque cards can be faked in front of a light so that the shape of the lighting/shadowing regions can be altered.

6.4.1   Trends and Analysis

Rendering engines are so powerful nowadays that getting the initial shadow properties set up should be very fast with the usual 3D modeling techniques (i.e., changing lights and objects can result in real-time feedback of the shadowing result). The cheats suggested by Barzel [36] should be able to get the precision of the shadow properties achieved quickly as well.

The inverse techniques described above might not prove as useful today, except in some very special cases. Work by Mitra and Pauly [414] and shadow art forms, as shown on the book cover by Kumi Yamashita and other work by Paul Pacotto [449] are such illustrations. It is not clear, however, if these inverse techniques would scale well to such difficult geometric constraints.

6.5   Final Words

In this chapter, some of the applications of shadow algorithms have been covered, including augmented reality, non-photorealistic environments, and shadows as interaction tools. While shadows play an important part of these applications, they often skip the shadow computations for different reasons (importance, performance, complexity, etc.). We are also certain there are other applications we have not covered (such as line-of-sight, though shadow algorithms seem overkill for such applications), but the ones covered in this chapter appear to be the most relevant to shadows.

CHAPTER 7

Conclusions

On the surface, shadow determination appears to be a simple task to achieve. However, as evidenced by the enormous set of research literature today and the trend that seems to continue, it is much more complex when the algorithms need to be robust, of good quality, and of good performance. We also doubt if there will ever be a single, feasible approach that unifies and serves all shadow needs. We strongly suspect that improvements will mostly come in the form of extensions to the three major shadow approaches—shadow volumes, shadow depth maps, and ray tracing.

From an application standpoint, the choice of categorized approach needs to be objectively determined, not chosen due to personal preference. The objective determinations can be evaluated in conjunction with

- Understanding the capabilities of the approach—see Section 1.3.

- Considerations for choosing an algorithm—see Section 1.5.

- Trends and analysis sections, including

  - Hard shadows—see Section 2.7.
  - Higher-order surfaces—see Section 3.2.4.
  - Image-based rendering for impostors—see Section 3.3.3.
  - Geometry images—see Section 3.4.1.
  - Point clouds—see Section 3.6.4.
  - Voxels—see Section 3.7.6.
  - Heightfields—see Section 3.8.3.
  - Soft shadows—see Section 4.9.
  - Bump mapping—see Section 5.2.2.
  - Advanced reflection models—see Section 5.3.1.
  - Semitransparent surfaces—see Section 5.4.4.

- ○ Highly complex thin materials—see Section 5.5.1.
- ○ Atmospheric shadows—see Section 5.6.4.
- ○ Motion blur—see Section 5.7.1.
- ○ Ambient occlusion—see Section 5.8.6.
- ○ Precomputed radiance transfer—see Section 5.9.1.
- ○ Global illumination—see Section 5.10.4.
- ○ Augmented reality—see Section 6.2.4.
- ○ Non-photorealistic environments—see Section 6.3.4.
- ○ Shadows as interaction tools—see Section 6.4.1.

From the experience of writing this book, we want to pass along our top ten reflections on shadow algorithms (other than how excited we are to see so much research work done on it), in no particular order:

1. Polygons are very important but not the only way to render shadows.

2. Don't ignore ray tracing as a potential solution for offline production and real-time shadow generation, even for today, and especially for tomorrow.

3. Self-shadowing (due to geometry or otherwise) is just as important for visual effects and feedback as is shadowing of other objects.

4. Don't ignore the terminator problem—it's real.

5. When using LOD techniques, ensure that there is some correlation of the LOD between the camera view and the light/shadow view; otherwise, bad self-shadowing may occur. And it may get worse (shifting or flickering arti-facts) over an animation with changing LODs.

6. Assuming that "objects can be closed" may not be a good assumption in many environments.

7. Something will go wrong, and usually at the worst time during production, so leave room for user input to "kludge" around the problem(s).

8. Beware of shadow algorithms with many or complicated special cases—once integrated into a larger system, the number or complexity of special cases may multiply.

9. Make sure the shadowing results look consistent spatially and temporally. Otherwise, even if it is easy to fool the human visual system, especially when the shadows represent supplementary visuals of the scene, some artifacts might catch your attention.

10. In a 1987 paper by Paul Heckbert [233] entitled "Ten Unsolved Problems in Rendering," number four was the lack of an efficient and robust approach to deal with shadows. While it is not a solved problem, it has certainly come a long way since then, with many more alternatives.

We hope you enjoyed the shadowy treatment of this book.

Bibliography

[1] A. Adamson and M. Alexa. "Ray Tracing Point Set Surfaces." In *Proceedings of Shape Modeling International*, pp. 272–279, 2003.

[2] M. Agrawala, R. Ramamoorthi, A. Heirich, and L. Moll. "Efficient Image-Based Methods for Rendering Soft Shadows." In *Proceedings of SIGGRAPH '00, Computer Graphics Proceedings, Annual Conference Series*, pp. 375–384. New York: ACM, 2000.

[3] T. Aila and T. Akenine-Möller. "A Hierarchical Shadow Volume Algorithm." In *Proceedings of Graphics Hardware*, pp. 15–23. New York: ACM, 2004.

[4] T. Aila and S. Laine. "Alias-Free Shadow Maps." In *Eurographics Symposium on Rendering*, pp. 161–166. Aire-la-Ville, Switzerland: Eurographics Association, 2004.

[5] T. Akenine-Möller and U. Assarsson. "Approximate Soft Shadows on Arbitrary Surfaces Using Penumbra Wedges." In *Eurographics Workshop on Rendering*, pp. 309–318. London: Springer-Verlag, 2002.

[6] T. Akenine-Möller and U. Assarsson. "On the Degree of Vertices in a Shadow Volume Silhouette." *Journal of Graphics, GPU, and Game Tools* 8:4 (2003), 21–24.

[7] T. Akenine-Möller, E. Haines, and N. Hoffman. *Real-Time Rendering*, Third edition. Natick, MA: A K Peters, 2008.

[8] G. Aldridge. "Robust, Geometry-Independent Shadow Volumes." In *Proceedings of Computer Graphics and Interactive Techniques in Australasia and Sout East Asia (GRAPHITE)*, pp. 250–253. New York: ACM, 2004.

[9] J. Amanatides and A. Woo. "A Fast Voxel Traversal Algorithm for Ray Tracing." *Proceedings of Eurographics* 1:3 (1987), 1–10.

[10] J. Amanatides. "Ray Tracing with Cones." *Proceedings of SIGGRAPH '84, Computer Graphics* 18:3 (1984), 129–136.

[11] T. Annen, T. Mertens, P. Bekaert, H. P. Seidel, and J. Kautz. "Convolution Shadow Maps." In *Eurographics Symposium on Rendering*, pp. 51–60. Aire-la-Ville, Switzerland: Eurographics Association, 2007.

[12] T. Annen, Z. Dong, T. Mertens, P. Bekaert, H. P. Seidel, and J. Kautz. "Real-Time, All-Frequency Shadows in Dynamic Scenes." *ACM Transactions on Graphics (Proceedings of SIGGRAPH '08)* 27:3 (2008), 34:1–8.

[13] T. Annen, T. Mertens, H. P. Seidel, E. Flerackers, and J. Kautz. "Exponential Shadow Maps." In *Proceedings of Graphics Interface*, pp. 155–161. Toronto, Canada: Canadian Information Processing Society, 2008.

[14] A. Appel. "Some Techniques for Shading Machine Rendering of Solids." In *Proceedings of AFIPS Spring Joint Computer Conference*, pp. 37–45. New York: ACM, 1968.

[15] J. Arvo and M. Hirvikorpi. "Compressed Shadow Maps." *The Visual Computer* 21:3 (2005), 125–138.

[16] J. Arvo and J. Westerholm. "Hardware Accelerated Soft Shadows Using Penumbra Quads." *Journal of WSCG* 12:1–3 (2004), 11–18.

[17] J. Arvo, M. Hirvikorpi, and J. Tyystjarvi. "Approximate Soft Shadows with an Image-Space Flood-Fill Algorithm." *Computer Graphics Forum (Proceedings of Eurographics)* 23:3 (2004), 271–280.

[18] J. Arvo. "Backward Ray Tracing." In *SIGGRAPH (Courses): Developments in Ray Tracing*, 1986.

[19] J. Arvo. "The Irradiance Jacobian for Partially Occluded Polyhedral Sources." In *Proceedings of SIGGRAPH '94, Computer Graphics Proceedings, Annual Conference Series*, pp. 343–350. New York: ACM, 1994.

[20] J. Arvo. "Tiled Shadow Maps." In *Proceedings of Computer Graphics International*, pp. 240–246. Washington, DC: IEEE Computer Society, 2004.

[21] J. Arvo. "Alias-Free Shadow Maps Using Graphics Hardware." *Journal of Graphics, GPU, and Game Tools* 12:1 (2007), 47–59.

[22] M. Ashikhmin, S. Premoze, and P.S. Shirley. "A Microfacet-Based BRDF Generator." In *Proceedings of SIGGRAPH '00, Computer Graphics Proceedings, Annual Conference Series*, pp. 65–74. New York: ACM, 2000.

[23] U. Assarsson and T. Akenine-Möller. "A Geometry-Based Soft Shadow Volume Algorithm Using Graphics Hardware." *ACM Transactions on Graphics (Proceedings of SIGGRAPH '03)* 22:3 (2003), 511–520.

[24] U. Assarsson and T. Akenine-Möller. "Occlusion Culling and Z-Fail for Soft Shadow Volume Algorithms." *The Visual Computer* 20:8–9 (2004), 601–612.

[25] U. Assarsson, M. Dougherty, M. Mounier, and T. Akenine-Möller. "An Optimized Soft Shadow Volume Algorithm with Real-Time Performance." In *Proceedings of Graphics Hardware*, pp. 33–40. New York: ACM, 2003.

[26] U. Assarsson. "A Real-Time Soft Shadow Volume Algorithm." PhD diss., Department of Computer Engineering, Chalmers University of Technology, 2003.

[27] B. Aszodi and L. Szirmay-Kalos. "Real-Time Soft Shadows with Shadow Accumulation." In *Eurographics (Short Papers)*, pp. 53–56. Aire-la-Ville, Switzerland: Eurographics Association, 2006.

[28] P. Atherton, K. Weiler, and D. Greenberg. "Polygon Shadow Generation." *Proceedings of SIGGRAPH '78* 12:3 (1978), 275–281.

[29] L. Atty, N. Holzschuch, M. Lapierre, J. Hasenfratz, C. Hansen, and F. Sillion. "Soft Shadow Maps: Efficient Sampling of Light Source Visibility." *Computer Graphics Forum* 25:4 (2006), 725–741.

[30] Y. Ayatsuka, S. Matsuoka, and J. Rekimoto. "Penumbra for 3D Interactions." In *Proceedings of ACM Symposium on User Interface Software and Technology*, pp. 165–166. New York: ACM, 1996.

[31] R. Azuma. "A Survey of Augmented Reality." *Teleoperators and Virtual Environments* 6:4 (1997), 355–385.

[32] K. Bala, B. Walter, and D. Greenberg. "Combining Edges and Points for Interactive High-Quality Rendering." *ACM Transactions on Graphics (Proceedings of SIGGRAPH '03)* 22:3 (2003), 631–640.

[33] H. Bao, J. Ying, and Q. Peng. "Shading with Curve Light Sources." *Computer Graphics Forum (Proceedings of Eurographics)* 14:3 (1995), 217–227.

[34] I. Baran, J. Chen, J. Ragan-Kelley, F. Durand, and J. Lehtinen. "A Hierarchical Volumetric Shadow Algorithm for Single Scattering." *ACM Transactions on Graphics (Proceedings of SIGGRAPH Asia '10)* 29:6.

[35] V. Barroso and W. Celes. "Improved Real-Time Shadow Mapping for CAD Models." In *Proceedings of Brazilian Symposium on Computer Graphics and Image Processing*, pp. 139–146. Washington, DC: IEEE Computer Society, 2007.

[36] R. Barzel. "Lighting Controls for Computer Cinematography." *Journal of Graphics, GPU, and Game Tools* 2:1 (1997), 1–20.

[37] H. Batagelo and I. Junior. "Real-Time Shadow Generation Using BSP Trees and Stencil Buffers." In *Proceedings of SIBGRAPI*, pp. 93–102. Washington, DC: IEEE Computer Society, 1999.

[38] L. Bavoil and M. Sainz. "Screen Space Ambient Occlusion." Technical report, NVIDIA, 2008.

[39] L. Bavoil and M. Sainz. "Multi-Layer Dual-Resolution Screen-Space Ambient Occlusion." In *SIGGRAPH Talks*, 2009.

[40] L. Bavoil and C. Silva. "Real-Time Soft Shadows with Cone Culling." In *SIGGRAPH (Sketches)*. New York: ACM, 2006.

[41] L. Bavoil, S. Callahan, and C. Silva. "Robust Soft Shadow Mapping with Depth Peeling." Technical Report UUSCI-2006-028, SCI Institute, University of Utah, 2006.

[42] L. Bavoil, M. Sainz, and R. Dimitrov. "Image-Space Horizon-Based Ambient Occlusion." In *SIGGRAPH Talks*, 2008.

[43] L. Bavoil. "Advanced Soft Shadow Mapping Techniques." Presented at Game Developers Conference, 2008.

[44] U. Behrens and R. Ratering. "Adding Shadows to a Texture-Based Volume Renderer." In *Proceedings of IEEE Symposium on Volume Visualization*, pp. 39–46. Washington, DC: IEEE Computer Society, 1998.

[45] C. Bentin and I. Wald. "Efficient Ray Traced Soft Shadows Using Multi-Frusta Tracing." In *Proceedings of High Performance Graphics*, pp. 135–144. New York: ACM, 2009.

[46] P. Bergeron. "A General Version of Crow's Shadow Volumes." *IEEE Computer Graphics and Applications* 6:9 (1986), 17–28.

[47] F. Bertails, C. Ménier, and M. Cani. "A Practical Self-Shadowing Algorithm for Interactive Hair Animation." In *Proceedings of Graphics Interface*, pp. 71–78. Toronto, Canada: Canadian Information Processing Society, 2005.

[48] J. Bestimt and B. Freitag. "Real-Time Shadow Casting Using Shadow Volumes." Technical report, Intel Corporation, 2002.

[49] M. Billeter, E. Sintorn, and U. Assarsson. "Real Time Volumetric Shadows Using Polygonal Light Volumes." In *Proceedings of High Performance Graphics*, pp. 39–45. New York: ACM, 2010.

[50] B. Bilodeau and M. Songy. "Real Time Shadows." In *Creativity '99 (Creative Labs Inc. Sponsored Game Developers Conference)*, 1999.

[51] V. Biri, A. Didier, and M. Sylvain. "Real Time Rendering of Atmospheric Scattering and Volumetric Shadows." *Journal of WSCG* 14:1–3 (2006), 65–72.

[52] J. Bittner, O. Mattausch, A. Silvennoinen, and M. Wimmer. "Shadow Caster Culling for Efficient Shadow Mapping." In *Proceedings of ACM Symposium on Interactive 3D Graphics and Games*, pp. 81–88. New York: ACM, 2011.

[53] J. Blinn. "Models of Light Reflection for Computer Synthesized Pictures." *Proceedings of SIGGRAPH '77* 11:2 (1977), 192–198.

[54] J. Blinn. "Simulation of Wrinkled Surfaces." *Proceedings of SIGGRAPH '78* 12:3 (1978), 286–292.

[55] J. Blinn. "A Generalization of Algebraic Surface Drawing." *ACM Transactions on Graphics* 1:3 (1982), 235–256.

[56] J. Blinn. "Light Reflection Functions for Simulation of Clouds and Dusty Surfaces." *Proceedings of SIGGRAPH '82, Computer Graphics* 16:3 (1982), 21–29.

[57] J. Blinn. "Jim Blinn's Corner: Me and My (Fake) Shadow." *IEEE Computer Graphics and Applications* 8:1 (1988), 82–86.

[58] J. Bloomenthal and K. Ferguson. "Polygonization of Non-Manifold Implicit Surfaces." In *Proceedings of SIGGRAPH '95, Computer Graphics Proceedings, Annual Conference Series*, pp. 309–316. New York: ACM, 1995.

[59] J. Bloomenthal. "Polygonization of Implicit Surfaces." *Computer Aided Geometric Design* 5:4 (1988), 341–355.

[60] M. Boehl and W. Stuerzlinger. "Real-Time Computation of Area Shadows—A Geometrical Algorithm." In *International Conference on Computer Graphics and Virtual Reality*, pp. 118–124. Providence, RI: CSREA, 2006.

[61] W. De Boer. "Smooth Penumbra Transitions with Shadow Maps." *Journal of Graphics, GPU, and Game Tools* 11:2 (2006), 59–71.

[62] M. Botsch, A. Hornung, M. Zwicker, and L. Kobbelt. "High-Quality Surface Splatting on Today's GPU." In *Proceedings of Eurographics Symposium on Point-Based Graphics*. Los Alamitos, CA: IEEE, 2005.

[63] W. Bouknight and K. Kelley. "An Algorithm for Producing Half-Tone Computer Graphics Presentations Shadows and Movable Light Sources." In *Proceedings of AFIPS Spring Joint Computer Conference*, 36, 36, pp. 1–10. New York: ACM, 1970.

[64] S. Boulos, D. Edwards, J. Lacewell, J. Kniss, J. Kautz, P. Shirley, and I. Wald. "Packet-Based Whitted and Distribution Ray Tracing." In *Proceedings of Graphics Interface*, pp. 177–184. Toronto, Canada: Canadian Information Processing Society, 2008.

[65] S. Brabec and H. P. Seidel. "Hardware-Accelerated Rendering of Anti-aliased Shadows with Shadow Maps." In *Proceedings of Computer Graphics International*, pp. 209–214. Washington, DC: IEEE Computer Society, 2001.

[66] S. Brabec and H. P. Seidel. "Single Sample Soft Shadows Using Depth Maps." In *Proceedings of Graphics Interface*, pp. 219–228. Toronto, Canada: Canadian Information Processing Society, 2002.

[67] S. Brabec and H. P. Seidel. "Shadow Volumes on Programmable Graphics Hardware." *Computer Graphics Forum (Proceedings of Eurographics)* 22:3 (2003), 433–440.

[68] S. Brabec, T. Annen, and H. P. Seidel. "Shadow Mapping for Hemispherical and Omnidirectional Light Sources." In *Proceedings of Computer Graphics International*, pp. 397–408. Washington, DC: IEEE Computer Society, 2002.

[69] S. Brabec, T. Annen, and H. P. Seidel. "Practical Shadow Mapping." *Journal of Graphics, GPU, and Game Tools* 7:4 (2003), 9–18.

[70] R. Brecheisen, B. Platel, A. Bartroli, and B. ter Haar Romenij. "Flexible GPU-Based Multi-Volume Ray-Casting." In *Proceedings of Vision, Modeling and Visualization*, pp. 1–6. Fairfax, VA, IOS Press, 2008.

[71] L. Brotman and N. Badler. "Generating Soft Shadows with a Depth Buffer Algorithm." *IEEE Computer Graphics and Applications* 4:10 (1984), 71–81.

[72] H. Buchholz, J. Dollner, M. Nienhaus, and F. Kirsch. "Real-Time Non-Photorealistic Rendering of 3D City Models." In *Workshop on Next Generation 3D City Models*. EuroSDR, 2005.

[73] B. Budge, T. Bernardin, J. Stuart, S. Sengupta, K. Joy, and J. Owens. "Out-of-Core Data Management for Path Tracing on Hybrid Resources." *Computer Graphics Forum (Proceedings of Eurographics)* 28:2 (2009), 385–396.

[74] T. Bujnak. "Extended Shadow Maps." In *Proceedings of the Central European Seminar on Computer Graphics (CESCG)*, 2004. Available online (http://www.cescg.org/CESCG-2004/papers/61_BujnakTomas.pdf).

[75] M. Bunnell and F. Pellacini. "Shadow Map Antialiasing." In *GPU Gems, Programming Techniques, Tips, and Tricks for Real-Time Graphics*, pp. 185–192. Reading, MA: Addison Wesley, 2004.

[76] M. Bunnell. "Dynamic Ambient Occlusion and Indirect Lighting." In *GPU Gems 2*, pp. 223–233. Reading, MA: Addison Wesley, 2005.

[77] B. Cabral, N. Max, and R. Springmeyer. "Bidirectional Reflection Functions from Surface Bump Maps." *Proceedings of SIGGRAPH '87, Computer Graphics* 21:4 (1987), 273–282.

[78] B. Cabral, N. Cam, and J. Foran. "Accelerated Volume Rendering and Tomographic Reconstruction Using Texture Mapping Hardware." In *Proceedings of the Symposium on Volume Visualization*, pp. 91–98. New York: ACM, 1994.

[79] X. Cai, Y. Jia, X. Wang, X. Hu, and R. Martin. "Rendering Soft Shadows Using Multi-layered Shadow Fins." *Computer Graphics Forum* 25:1 (2006), 15–28.

[80] A. Campbell and D. Fussell. "Adaptive Mesh Generation for Global Diffuse Illumination." *Proceedings of SIGGRAPH '90, Computer Graphics* 24:4 (1990), 155–164.

[81] J. Carmack. "John Carmack on Shadow Volumes.", 2000. Personal Communication to Mark Kilgard, http://www.cnblogs.com/Pointer/archive/2004/07/28/28036.html.

[82] L. Carpenter. "The A-buffer, an Antialiased Hidden Surface Method." *Proceedings of SIGGRAPH '84, Computer Graphics* 18:3 (1984), 103–108.

[83] E. Chan and F. Durand. "Rendering Fake Soft Shadows with Smoothies." In *Eurographics Symposium on Rendering*, pp. 208–218. Aire-la-Ville, Switzerland: Eurographics Association, 2003.

[84] E. Chan and F. Durand. "An Efficient Hybrid Shadow Rendering Algorithm." In *Eurographics Symposium on Rendering*, pp. 185–195. Aire-la-Ville, Switzerland: Eurographics Association, 2004.

[85] C. Chang, B. Lin, Y Chen, and Y. Chiu. "Real-Time Soft Shadow for Displacement Mapped Surfaces." In *Multimedia and Expo*, pp. 1254–1257. Piscataway, NJ: IEEE Press, 2009.

[86] S. Chattopadhyay and A. Fujimoto. "Bi-directional Ray Tracing." In *Proceedings of Computer Graphics International*, pp. 335–343. Washington, DC: IEEE Computer Society, 1987.

[87] B. Chen and M. Nguyen. "POP: A Hybrid Point and Polygon Rendering System for Large Data." In *Proceedings of IEEE Visualization*, pp. 45–52. Washington, DC: IEEE Computer Society, 2001.

[88] S. Chen and L. Williams. "View Interpolation for Image Synthesis." In *Proceedings of SIGGRAPH '93, Computer Graphics Proceedings, Annual Conference Series*, pp. 279–288. New York: ACM, 1993.

[89] X. Chen and Y.H. Yang. "Recovering Stereo Depth Maps Using Single Gaussian Blurred Color Structured Light Pattern." Technical Report TR11-07, Department of Computing Science, University of Alberta, 2011.

[90] J. Chen, I. Baran, F. Durand, and J. Jarosz. "Real-Time Volumetric Shadows Using 1D Min-max Mipmaps." In *Proceedings of ACM Symposium on Interactive 3D Graphics and Games*, pp. 39–46. New York: ACM, 2011.

[91] N. Chin and S. Feiner. "Near Real-Time Object-Precision Generation Using BSP Trees." *Proceedings of SIGGRAPH '89, Computer Graphics* 23:3 (1989), 99–106.

[92] N. Chin and S. Feiner. "Fast Object-Precision Shadow Generation for Area Light Sources Using BSP Trees." In *Proceedings of ACM Symposium on Interactive 3D Graphics*, pp. 21–30. New York: ACM, 1992.

[93] Y. Chio, J. Choi, D. Kim, and K. Yoon. "Shadow Generation in Virtual Reality Space Using Shadow Polygons." *Proceeding of KISS* 24:1 (1997), 419–422.

[94] H. Choi and C. Kyung. "PYSHA—A Shadow Testing Acceleration Scheme for Ray Tracing." *Computer Aided Design* 24:2 (1992), 93–104.

[95] H. Chong and S. Gortler. "A Lixel for Every Pixel." In *Eurographics Symposium on Rendering*, pp. 167–172. Aire-la-Ville, Switzerland: Eurographics Association, 2004.

[96] H. Chong and S. Gortler. "Scene Optimized Shadow Mapping." Technical Report TR-11-06, Harvard Computer Science, 2006.

[97] H. Chong. "Real-Time Perspective Optimal Shadow Maps." BA thesis, Computer Science, Harvard College, Cambridge, Massachusetts, 2003.

[98] P. Christensen. "Global Illumination and All That." In *SIGGRAPH (Courses): Course 9*, 2003.

[99] Y. Chrysanthou and M. Slater. "Dynamic Changes to Scenes Represented as BSP Trees." *Computer Graphics Forum (Proceedings of Eurographics)* 11:3 (1992), 321–332.

[100] Y. Chrysanthou and Mel Slater. "Shadow Volume BSP Trees for Computation of Shadows in Dynamic Scenes." In *Proceedings of ACM Symposium on Interactive 3D Graphics*, pp. 45–50. New York: ACM, 1995.

[101] Y. Chrysanthou and M. Slater. "Incremental Updates to Scenes Illuminated by Area Light Sources." In *Eurographics Workshop on Rendering*, pp. 103–114. London: Springer-Verlag, 1997.

[102] Y. Chuang, D. Goldman, B. Curless, D. Salesin, and R. Szeliski. "Shadow Matting and Compositing." *ACM Transactions on Graphics (Proceedings of SIGGRAPH '03)* 22:3 (2003), 494–500.

[103] M. Cohen and J. Wallace. *Radiosity and Realistic Image Synthesis*. New York: Academic Press, 1993.

[104] M. Cohen, S. Chen, J. Wallace, and D. Greenberg. "A Progressive Refinement Approach to Fast Radiosity Image Generation." *Proceedings of SIGGRAPH '88, Computer Graphics* 22:4 (1988), 75–84.

[105] D. Cohen-Or, E. Rich, U. Lerner, and V. Shenkar. "A Real-Time Photo-Realistic Visual Flythrough." In *IEEE Transactions on Visualization and Computer Graphics*, pp. 255–265. Washington, DC: IEEE Computer Society, 1996.

[106] D. Cohen-Or, Y. Chrysanthou, C. Silva, and F. Durand. "A Survey of Visibility for Walkthrough Applications." *IEEE Transactions on Visualization and Computer Graphics* 9:3 (2003), 412–431.

[107] M. Contreras, A. Valadez, and A. Martinez. "Dual Sphere-Unfolding Method for Single Pass Omni-Directional Shadow Mapping." In *SIGGRAPH (Sketches)*. New York: ACM, 2011.

[108] R. Cook and L. Carpenter. "Distributed Ray Tracing." *Proceedings of SIGGRAPH '84, Computer Graphics* 18:3 (1984), 109–115.

[109] R. Cook and K. Torrance. "A Reflectance Model for Computer Graphics." *Proceedings of SIGGRAPH '81* 15:3 (1981), 301–316.

[110] R. Cook, L. Carpenter, and E. Catmull. "The Reyes Image Rendering Architecture." *Proceedings of SIGGRAPH '87, Computer Graphics* 21:4 (1987), 95–102.

[111] S. Coquillart and M. Gangnet. "Shaded Display of Digital Maps." In *IEEE Computer Graphics and Applications*, pp. 35–42. Washington, DC: IEEE Computer Society, 1984.

[112] C. Crassin, F. Neyret, S. Lefebvre, and E. Eisemann. "GigaVoxels: Ray-guided Streaming for Efficient and Detailed Voxel Rendering." In *Proceedings of ACM Symposium on Interactive 3D Graphics and Games*, pp. 15–22. New York: ACM, 2009.

[113] C. Crassin, F. Neyret, S. Lefebvre, M. Sainz, and E. Eisemann. "Beyond Triangles: GigaVoxels Effects in Video Games." In *SIGGRAPH (Sketches)*. New York: ACM, 2010.

[114] F. Crow. "Shadow Algorithms for Computer Graphics." In *Proceedings of SIGGRAPH '77*, pp. 242–248. New York: ACM, 1977.

[115] C. Dachsbacher and M. Stamminger. "Translucent Shadow Maps." In *Eurographics Symposium on Rendering*, pp. 197–201. Aire-la-Ville, Switzerland: Eurographics Association, 2003.

[116] C. Dachsbacher and M. Stamminger. "Splatting Indirect Illumination." In *Proceedings of ACM Symposium on Interactive 3D Graphics and Games*, pp. 93–100. New York: ACM, 2006.

[117] Q. Dai, B. Yang, and J. Feng. "Reconstructable Geometry Shadow Maps." In *Proceedings of ACM Symposium on Interactive 3D Graphics and Games*, pp. 4:1–1. New York: ACM, 2008.

[118] S. Danner and C. Winklhofer. "Cartoon Style Rendering." Technical report, TU Wien, 2008.

[119] P. Debevec. "Rendering Synthetic Objects into Real Scenes: Bridging Traditional and Image-Based Graphics with Global Illumination and High Dynamic Range Photography." In *Proceedings of SIGGRAPH '98, Computer Graphics Proceedings, Annual Conference Series*, pp. 189–198. New York: ACM, 1998.

[120] P. Debevec. "A Median Cut Algorithm for Light Probe Sampling." In *Computer Graphics and Interactive Techniques*. New York: ACM, 2005.

[121] P. Decaudin and F. Neyret. "Volumetric Billboards." *Computer Graphics Forum* 28:8 (2009), 2079–2089.

[122] P. Decaudin. "Cartoon Looking Rendering of 3D Scenes." Technical report, INRIA, 1996.

[123] X. Décoret, F. Durand, F. Sillion, and J. Dorsey. "Billboard Clouds for Extreme Model Simplification." *ACM Transactions on Graphics (Proceedings of SIGGRAPH '03)* 22:3 (2003), 689–696.

[124] C. DeCoro and S. Rusinkiewicz. "Subtractive Shadows: A Flexible Framework for Shadow Level-of-Detail." Technical Report TR-781-07, Princeton University, 2007.

[125] C. DeCoro, F. Cole, A. Finkelstein, and S. Rusinkiewicz. "Stylized Shadows." In *Proceedings of Symposium on Non-Photorealistic Animation and Rendering*, pp. 77–83. New York: ACM, 2007.

[126] J. Demouth and X. Goaoc. "Computing Direct Shadows Cast by Convex Polyhedra." In *Proceedings of European Symposium on Computational Geometry*, 2009. Available online (http://2009.eurocg.org/abstracts.pdf).

[127] J. Demouth, O. Devillers, H. Everett, M. Glisse, S. Lazard, and R. Seidel. "Between Umbra and Penumbra." In *Proceedings of Symposium on Computational Geometry*, pp. 265–274. New York: ACM, 2007.

[128] P. Diefenbach and N. Badler. "Pipeline Rendering: Interactive Refractions, Reflections and Shadows." *Displays* 15:3 (1994), 173–180.

[129] D. Dietrich. "Practical Priority Buffer Shadows." In *Game Programming Gems II*, pp. 481–487. Hingham, MA: Charles River Media, 2001.

[130] E. Diktas and A. Sahiner. "Parallel ID Shadow-Map Decompression on GPU." In *Proceedings of Symposium on Parallel and Distributed Computing*, pp. 79–84. Washington, DC: IEEE Computer Society, 2010.

[131] R. Dimitrov. "Cascaded Shadow Maps." Technical report, NVIDIA Corporation, 2007.

[132] Q. Dinh, R.A. Metoyer, and G. Turk. "Real-Time Lighting Changes for Image-Based Rendering." In *Proceedings of the IASTED International Conference, Computer Graphics and Imaging*, pp. 58–63. Calgary, Canada: ACTA Press, 1998.

[133] P. Djeu, S. Keely, and W. Hunt. "Accelerating Shadow Rays Using Volumetric Occluders and Modified kd-tree Traversal." In *Proceedings of High Performance Graphics*, pp. 69–76. New York: ACM, 2009.

[134] Y. Dobashi, T. Yamamoto, and T. Nishita. "Interactive Rendering Method for Displaying Shafts of Light." In *Proceedings of Pacific Graphics*, pp. 31–37. Washington, DC: IEEE Computer Society, 2000.

[135] Y. Dobashi, T. Nishita, and T. Yamamoto. "Interactive Rendering of Atmospheric Scattering Effects Using Graphics Hardware." In *Proceedings of Graphics Hardware*, pp. 99–108. Aire-la-Ville, Switzerland: Eurographics Association, 2002.

[136] S. Dobbyn, J. Hamill, K. O'Conor, and C. O'Sullivan. "Geopostors: A Real-Time Geometry / Impostor Crowd Rendering System." In *Proceedings of ACM Symposium on Interactive 3D Graphics and Games*, pp. 95–102. New York: ACM, 2005.

[137] P. Dobrev, P. Rosenthal, and L. Linsen. "Interactive Image-Space Point Cloud Rendering with Transparency and Shadows." In *Computer Graphics, Visualization, and Computer Vision*, edited by Vaclav Skala. pp. 101–108, Union Agency—Science Press, 2010.

[138] Z. Dong and B. Yang. "Variance Soft Shadow Mapping." In *Proceedings of ACM Symposium on Interactive 3D Graphics and Games*, pp. 18:1–1. New York: ACM, 2010.

[139] W. Donnelly and J. Demers. "Generating Soft Shadows Using Occlusion Interval Maps." In *GPU Gems, Programming Techniques, Tips, and Tricks for Real-Time Graphics*, pp. 205–215. Reading, MA: Addison Wesley, 2004.

[140] W. Donnelly and A. Lauritzen. "Variance Shadow Maps." In *Proceedings of ACM Symposium on Interactive 3D Graphics and Games*, pp. 161–165. New York: ACM, 2006.

[141] G. Drettakis and E. Fiume. "A Fast Shadow Algorithm for Area Light Sources Using Backprojection." In *Proceedings of SIGGRAPH '94, Computer Graphics Proceedings, Annual Conference Series*, pp. 223–230. New York: ACM, 1994.

[142] G. Drettakis and E. Fiume. "Structured Penumbral Irradiance Computation." *IEEE Transactions on Visualization and Computer Graphics* 4:2 (1996), 299–312.

[143] G. Drettakis, L. Robert, and S. Gougnoux. "Interactive Common Illumination for Computer Augmented Reality." In *Eurographics Workshop on Rendering*, pp. 45–56. London: Springer-Verlag, 1997.

[144] F. Duguet and G. Drettakis. "Robust Epsilon Visibility." *ACM Transactions on Graphics (Proceedings of SIGGRAPH '02)* 21:3 (2002), 567–575.

[145] F. Duguet and G. Drettakis. "Flexible Point-Based Rendering on Mobile Devices." In *IEEE Computer Graphics and Applications*, pp. 57–63. Washington, DC: IEEE Computer Society, 2004.

[146] F. Duguet. "Shadow Computations Using Robust Epsilon Visibility." PhD diss., INRIA, 2004.

[147] F. Durand, G. Drettakis, and C. Puech. "The Visibility Skeleton: A Powerful and Efficient Multi-purpose Global Visibility Tool." In *Proceedings of SIGGRAPH '97, Computer Graphics Proceedings, Annual Conference Series*, pp. 89–100. New York: ACM, 1997.

[148] F. Durand, N. Holzschuch, C. Soler, E. Chan, and F. Sillion. "A Frequency Analysis of Light Transport." *ACM Transactions on Graphics (Proceedings of SIGGRAPH '05)* 24:3 (2005), 1115–1126.

[149] F. Durand. "3D Visibility: Analytical Study and Applications." PhD diss., Université Joseph Fourier, Grenoble, France, 1999.

[150] P. Dutré, P. Tole, and D. Greenberg. "Approximate Visibility for Illumination Computations Using Point Clouds." Technical Report PCG-00-1, Cornell University, 2000.

[151] P. Dutré, K. Bala, and P. Bekaert. *Advanced Global Illumination*, Second edition. Natick, MA: A K Peters, 2006.

[152] D. Ebert and R. Parent. "Rendering and Animation of Gaseous Phenomena by Combining Fast Volume and Scanline A-buffer Techniques." *Proceedings of SIGGRAPH '90, Computer Graphics* 24:4 (1990), 357–366.

[153] K. Egan, F. Durand, and R. Ramamoorthi. "Practical Filtering for Efficient Ray-Traced Directional Occlusion." *ACM Transactions on Graphics (Proceedings of SIGGRAPH Asia '11)* 30:6 (2011), 180:1–10.

[154] K. Egan, F. Hecht, F. Durand, and R. Ramamoorthi. "Frequency Analysis and Sheared Filtering for Shadow Light Fields of Complex Occluders." *ACM Transactions on Graphics (Proceedings of SIGGRAPH '11)* 30:3 (2011), 1–13.

[155] E. Eisemann and X. Décoret. "Fast Scene Voxelization and Applications." In *Proceedings of ACM Symposium on Interactive 3D Graphics and Games*, pp. 71–78. New York: ACM, 2006.

[156] E. Eisemann and X. Décoret. "Plausible Image Based Soft Shadows Using Occlusion Textures." In *Proceedings of Brazilian Symposium on Computer Graphics and Image Processing*, pp. 155–162. Washington, DC: IEEE Computer Society, 2006.

[157] E. Eisemann and X. Décoret. "Occlusion Textures for Plausible Soft Shadows." *Computer Graphics Forum* 27:1 (2008), 13–23.

[158] E. Eisemann, M. Schwartz, U. Assarsson, and M. Wimmer. *Real-Time Shadows*. Natick, MA: A K Peters, 2011.

[159] E. Enderton, E. Sintorn, P. Shirley, and D. Luebke. "Stochastic Transparency." In *Proceedings of ACM Symposium on Interactive 3D Graphics and Games*, pp. 157–164. New York: ACM, 2010.

[160] K. Engel, M. Hadwiger J. Kniss, C. Rezk-Salama, and D. Weiskopf. *Real-Time Volume Graphics*. Natick, MA: A K Peters, 2006.

[161] W. Engel. "Cascaded Shadow Maps." In *ShaderX⁵: Advanced Rendering Techniques*, pp. 197–206. Hingham, MA: Charles River Media, 2006.

[162] T. Engelhardt and C. Dachsbacher. "Epipolar Sampling for Shadows and Crepuscular Rays in Participating Media with Single Scattering." In *Proceedings of ACM Symposium on Interactive 3D Graphics and Games*, pp. 119–125. New York: ACM, 2010.

[163] A. Entezari, R. Scoggins, T. Möller, and R. Machiraju. "Shading for Fourier Volume Rendering." In *Proceedings of IEEE Symposium on Volume Visualization and Graphics*. Washington, DC: IEEE Computer Society, 2002.

[164] D. Eo and C. Kyung. "Hybrid Shadow Testing Scheme for Ray Tracing." *Computer Aided Design* 21:1 (1989), 38–48.

[165] C. Everitt and M. Kilgard. "Practical and Robust Stenciled Shadow Volumes for Hardware-Accelerated Rendering." Technical report, NVIDIA Corporation, 2002.

[166] C. Everitt, A. Rege, and C. Cebenoyan. "Hardware Shadow Mapping." Technical report, NVIDIA Corporation, 2001.

[167] C. Everitt. "Interactive Order-Independent Transparency." Technical report, NVIDIA Corporation, 2001.

[168] K. Fauerby and C. Kjaer. "Real-Time Soft Shadows in a Game Engine." Master's thesis, University of Aarhus, 2003.

[169] M. Fawad. "Investigating Shadow Volumes on Novel Paradigm." In *International Conference Graphicon*, 2006.

[170] M.S. Fawad. "LOD based Real Time Shadow Generation for Subdivison Surfaces." *International Journal of Computer and Electrical Engineering* 2:1 (2010), 170–174.

[171] S. Fernandez, K. Bala, and D. Greenberg. "Local Illumination Environments for Direct Lighting Acceleration." In *Eurographics Workshop on Rendering*, pp. 7–14. London: Springer-Verlag, 2002.

[172] R. Fernando, S. Fernandez, K. Bala, and D. Greenberg. "Adaptive Shadow Maps." In *Proceedings of SIGGRAPH '01, Computer Graphics Proceedings, Annual Conference Series*, pp. 387–389. New York: ACM, 2001.

[173] R. Fernando. "Percentage-Closer Soft Shadows." In *SIGGRAPH (Sketches)*. New York: ACM, 2005.

[174] B. Fleming. *3D Photorealism Toolkit*. New York: John Wiley and Sons, 1988.

[175] V. Forest, L. Barthe, and M. Paulin. "Realistic Soft Shadows by Penumbra-Wedges Blending." In *Proceedings of Graphics Hardware*, pp. 39–46. New York: ACM, 2006.

[176] V. Forest, L. Barthe, and M. Paulin. "Accurate Shadows by Depth Complexity Sampling." *Computer Graphics Forum (Proceedings of Eurographics)* 27:2 (2008), 663–674.

[177] V. Forest, L. Barthe, G. Guennebaud, and M. Paulin. "Soft Textured Shadow Volume." *Computer Graphics Forum (Eurographics Symposium on Rendering)* 28:4 (2009), 1111–1121.

[178] A. Formella and A. Lukaszewski. "Fast Penumbra Calculation in Ray Tracing." *Journal of WSCG* 6:1–3 (1998), 238–245.

[179] T. Forsyth. "Self-Shadowing Bumpmap Using 3D Texture Hardware." *Journal of Graphics, GPU, and Game Tools* 7:4 (2003), 19–26.

[180] A. Fournier and J. Buchanan. "Chebyshev Polynomials for Boxing and Intersections of Parametric Curves and Surfaces." *Computer Graphics Forum (Proceedings of Eurographics)* 13:3 (1994), 127–142.

[181] A. Fournier and D. Fussell. "On the Power of the Frame Buffer." *ACM Transactions on Graphics* 7:2 (1988), 103–128.

[182] A. Fournier, A. Gunawan, and C. Romanzin. "Common Illumination between Real and Computer Generated Scenes." In *Proceedings of Graphics Interface*, pp. 254–261. Toronto, Canada: Canadian Information Processing Society, 1993.

[183] M. Friedell and S. Kochhar. "Toward Reliable Polygon Set Operations." In *Proceedings of Graphics Interface*, pp. 167–174. Toronto, Canada: Canadian Information Processing Society, 1990.

[184] H. Fuchs, J. Goldfeather, J.P. Hultquist, S. Spach, J.D. Austin, F.P. Brooks Jr., J.G. Eyles, and J. Poulton. "Fast Spheres, Shadows, Textures, Transparencies, and Image Enhancements in Pixel-Planes." *Proceedings of SIGGRAPH '85, Computer Graphics* 19:3 (1985), 111–120.

[185] Mathieu Gauthier and Pierre Poulin. "Preserving Sharp Edges in Geometry Images." In *Proceedings of Graphics Interface*, pp. 1–6. Toronto, Canada: Canadian Information Processing Society, 2009.

[186] P. Gautron, J. Marvie, and G. François. "Volumetric Shadow Mapping." In *SIGGRAPH (Sketches)*. New York: ACM, 2009.

[187] S. Gebhardt, E. Payzer, and L. Salemann. "Polygons, Point-Clouds, and Voxels, a Comparison of High-Fidelity Terrain Representations." In *Simulation Interoperability Workshop and Special Workshop on Reuse of Environmental Data for Simulation— Processes, Standards, and Lessons Learned*, 2009.

[188] J. Genetti and D. Gordon. "Ray Tracing with Adaptive Supersampling in Object Space." In *Proceedings of Graphics Interface*, pp. 70–77. Toronto, Canada: Canadian Information Processing Society, 1993.

[189] J. Genetti, D. Gordon, and G. Williams. "Adaptive Supersampling in Object Space Using Pyramidal Rays." *Computer Graphics Forum* 17:1 (1998), 29–54.

[190] P. Gerasimov. "Omnidirectional Shadow Mapping." In *GPU Gems, Programming Techniques, Tips, and Tricks for Real-Time Graphics*, pp. 193–203. Reading, MA: Addison Wesley, 2004.

[191] S. Ghali, E. Fiume, and H. P. Seidel. "Object-Space, Connectivity-Preserving, Shadow Computation." In *Proceedings of Vision, Modeling and Visualization*, pp. 281–289. Fairfax, VA: IOS Press, 2000.

[192] S. Ghali, E. Fiume, and H. P. Seidel. "Shadow Computation: A Unified Perspective." In *Eurographics State-of-the-Art Reports*, pp. 1–17. Aire-la-Ville, Switzerland: Eurographics, 2000.

[193] D. Ghazanfarpour and J. Hasenfratz. "A Beam Tracing Method with Precise Anti-aliasing for Polyhedral Scenes." *Computers and Graphics* 22:1 (1998), 103–115.

[194] S. Gibson and A. Murta. "Interactive Rendering with Real World Illumination." In *Eurographics Workshop on Rendering*, pp. 365–376. London: Springer-Verlag, 2000.

[195] S. Gibson, J. Cook, T. Howard, and R. Hubbold. "Rapid Shadow Generation in Real-world Lighting Environments." In *Eurographics Symposium on Rendering*, pp. 219–229. Aire-la-Ville, Switzerland: Eurographics Association, 2003.

[196] M. Giegl and M. Wimmer. "Fitted Virtual Shadow Maps." In *Proceedings of Graphics Interface*, pp. 159–168. Toronto, Canada: Canadian Information Processing Society, 2007.

[197] M. Giegl and M. Wimmer. "Queried Virtual Shadow Maps." In *Proceedings of ACM Symposium on Interactive 3D Graphics and Games*, pp. 65–72. New York: ACM, 2007.

[198] A.S. Glassner, editor. *An Introduction to Ray Tracing.* New York: Academic Press, 1989.

[199] B. Gooch, P.-P.J. Sloan, A. Gooch, P. Shirley, and R. Riesenfeld. "Interactive Technical Illustration." In *Proceedings of ACM Symposium on Interactive 3D Graphics*, pp. 31–38. New York: ACM, 1999.

[200] C. Goral, K. Torrance, D. Greenberg, and B. Battaile. "Modelling the Interaction of Light between Diffuse Surfaces." *Proceedings of SIGGRAPH '84, Computer Graphics* 18:3 (1984), 213–222.

[201] N. Govindaraju, B. Lloyd, S. Yoon, A. Sud, and D. Manocha. "Interactive Shadow Generation in Complex Environments." *ACM Transactions on Graphics (Proceedings of SIGGRAPH '03)* 22:3 (2003), 501–510.

[202] C. Grant. "Visibility Algorithms in Image Synthesis." PhD diss., University of California, 1992.

[203] R. Green. "Spherical Harmonic Lighting: The Gritty Details." Presented at Game Developers Conference, 2003.

[204] T. Grosch, T. Eble, and S. Mueller. "Consistent Interactive Augmentation of Live Camera Images with Correct Near-Field Illumination." In *Proceedings of Symposium on Virtual Reality Software and Technology*, pp. 125–132. New York: ACM, 2007.

[205] J. Grossman and W. Dally. "Point Sample Rendering." In *Eurographics Workshop on Rendering*, pp. 181–192. London: Springer-Verlag, 1998.

[206] X. Gu, S.J. Gortler, and H. Hoppe. "Geometry Images." *ACM Transactions on Graphics (Proceedings of SIGGRAPH '02)* 21:3 (2002), 355–361.

[207] G. Guennebaud and M. Gross. "Algebraic Point Set Surfaces." *ACM Transactions on Graphics (Proceedings of SIGGRAPH '07)* 26:3 (2007), 23.

[208] G. Guennebaud, L. Barthe, and M. Paulin. "Real-Time Soft Shadow Mapping by Backprojection." In *Eurographics Symposium on Rendering*, pp. 227–234. Aire-la-Ville, Switzerland: Eurographics Association, 2006.

[209] G. Guennebaud, L. Barthe, and M. Paulin. "High-Quality Adaptive Soft Shadow Mapping." *Computer Graphics Forum (Proceedings of Eurographics)* 26:3 (2007), 525–534.

[210] J. Gumbau, M. Chover, and M. Sbert. "Screen Space Soft Shadows." In *GPU Pro*, pp. 477–491. Natick, MA: A K Peters, 2010.

[211] J. Gumbau, L. Szirmay-Kalos, M. Sbert, and M. Sells. "Improving Shadow Map Filtering with Statistical Analysis." In *Eurographics (Short Papers)*, pp. 33–36. Aire-la-Ville, Switzerland: Eurographics Association, 2011.

[212] Z. Gunjee, T. Fujimoto, and N. Chiba. "Point Splatting based on Translucent Shadow Mapping and Hierarchical Bucket Sorting." *Journal of the Society for Art and Science* 6:1 (2007), 21–36.

[213] J. Gunther, S. Popov, H. P. Seidel, and P. Slusallek. "Realtime Ray Tracing on GPU with BVH-Based Packet Traversal." In *Eurographics Symposium on Interactive Ray Tracing*, pp. 113–118. Washington, DC: IEEE Computer Society, 2007.

[214] V. Gusev. "Extended Perspective Shadow Maps (XPSM)." http://xpsm.org, 2007. Available online (http://citeseerx.ist.psu.edu/viewdoc/summary?doi=10.1.1.95.7159).

[215] M. Guthe, P. Borodin, A. Balazs, and R. Klein. "Real-Time Appearance Preserving Out-of-Core Rendering with Shadows." In *Eurographics Symposium on Rendering*, pp. 69–79. Aire-la-Ville, Switzerland: Eurographics Association, 2004.

[216] M. Guthe, A. Balazs, and R. Klein. "GPU-Based Trimming and Tessellation of NURBS and T-Spline Surfaces." *ACM Transactions on Graphics (Proceedings of SIGGRAPH '05)* 24:3 (2005), 1016–1023.

[217] M. Hadwiger, A. Kratz, C. Sigg, and K. Buhler. "GPU-Accelerated Deep Shadow Maps for Direct Volume Rendering." In *Proceedings of Graphics Hardware*, pp. 49–52. New York: ACM, 2006.

[218] P. Haeberli and K. Akeley. "The Accumulation Buffer: Hardware Support for High-Quality Rendering." *Proceedings of SIGGRAPH '90, Computer Graphics* 24:4 (1990), 309–318.

[219] E. Haines and D. Greenberg. "The Light Buffer: A Ray Tracer Shadow Testing Accelerator." *IEEE Computer Graphics and Applications* 6:9 (1986), 6–16.

[220] E. Haines. "Soft Planar Shadows Using Plateaus." *Journal of Graphics, GPU, and Game Tools* 6:1 (2001), 19–27.

[221] R. Hall and D. Greenberg. "A Testbed for Realistic Image Synthesis." *IEEE Computer Graphics and Applications* 3:8 (1983), 10–20.

[222] M. Haller, S. Drab, and W. Hartmann. "A Real-Time Shadow Approach for an Augmented Reality Application Using Shadow Volumes." In *Proceedings of ACM Symposium on Virtual Reality Software and Technology*, pp. 56–65. New York: ACM, 2003.

[223] A. Hallingstad. "Projective Shadows Using Stencil Buffer." http://www.geocities.com/arne9184/articlestutorials/projective_shadows_using_stencil_buffer.htm, 2002.

[224] P. Hanrahan. "Ray Tracing Algebraic Surfaces." *Proceedings of SIGGRAPH '83, Computer Graphics* 17:3 (1983), 83–89.

[225] J. Hart, P. Dutré, and D. Greenberg. "Direct Illumination with Lazy Visibility Evaluation." *ACM Transactions on Graphics (Proceedings of SIGGRAPH '02)* 21:3 (1999), 147–154.

[226] T. Harter, M. Osswald, and U. Assarsson. "Soft Shadow Volumes for Ray Tracing with Frustum Shooting." Technical report, Chalmers University of Technology, 2008.

[227] J. Hasenfratz, M. Lapierre, H. Holzschuch, and F. Sillion. "A Survey of Real-Time Soft Shadow Algorithms." *Computer Graphics Forum (Eurographics State-of-the-Art Reports)* 22:4 (2003), 753–774.

[228] Jon Hasselgren and Thomas Akenine-Möller. "Textured Shadow Volumes." *Journal of Graphics, GPU, and Game Tools* 12:4 (2007), 59–72.

[229] M. Hašan, F. Pellacini, and K. Bala. "Matrix Row-Column Sampling for the Many-Light Problem." *ACM Transactions on Graphics (Proceedings of SIGGRAPH '07)* 26:3 (2007), 26: 1–10.

[230] X. He, K. Torrance, F. Sillion, and D.P. Greenberg. "A Comprehensive Physical Model for Light Reflection." *Proceedings of SIGGRAPH '91, Computer Graphics* 25:4 (1991), 175–186.

[231] P. Heckbert and P. Hanrahan. "Beam Tracing Polygonal Objects." *Proceedings of SIGGRAPH '84, Computer Graphics* 18:3 (1984), 118–127.

[232] P. Heckbert and M. Herf. "Simulating Soft Shadows with Graphics Hardware." Technical Report CMU-CS-97-104, Carnegie Mellon University, 1997.

[233] P. Heckbert. "Ten Unsolved Problems in Rendering." Presented at the *Workshop on Rendering Algorithms and Systems, at Graphics Interface Conference*, 1987.

[234] P. Heckbert. "Discontinuity Meshing for Radiosity." In *Eurographics Workshop on Rendering*, pp. 203–216. London: Springer-Verlag, 1992.

[235] G. Heflin and G. Elber. "Shadow Volume Generation from Free Form Surfaces." In *Proceedings of Computer Graphics International*, pp. 115–126. Washington, DC: IEEE Computer Society, 1993.

[236] T. Heidmann. "Real Shadows Real Time." *IRIS Universe* 18 (1991), 28–31.

[237] W. Heidrich, S. Brabec, and H. P. Seidel. "Soft Shadow Maps for Linear Lights." In *Eurographics Workshop on Rendering*, pp. 269–280. London: Springer-Verlag, 2000.

[238] W. Heidrich, K. Daubert, J. Kautz, and H. P. Seidel. "Illuminating Micro Geometry based on Precomputed Visibility." In *Proceedings of SIGGRAPH '00, Computer Graphics Proceedings, Annual Conference Series*, pp. 455–464. New York: ACM, 2000.

[239] C. Henning and P. Stephenson. "Accelerating the Ray Tracing of Height Fields." In *Proceedings of Computer Graphics and Interactive Techniques in Australasia and Sout East Asia (GRAPHITE)*, pp. 254–258. New York: ACM, 2004.

[240] M. Herf and P. Heckbert. "Fast Soft Shadows." In *SIGGRAPH (Sketches)*. New York: ACM, 1996.

[241] K.P. Herndon, R.C. Zeleznik, D.C. Robbins, D.B. Conner, S.S. Snibbe, and A. van Dam. "Interactive Shadows." In *Proceedings of ACM Symposium on User Interface Software and Technology (UIST)*, pp. 1–6. New York: ACM, 1992.

[242] F. Hernell, P. Ljung, and A. Ynnerman. "Local Ambient Occlusion in Direct Volume Rendering." *IEEE Transactions on Visualization and Computer Graphics* 16:4 (2010), 548–559.

[243] S. Hertel, K. Hormann, and R. Westermann. "A Hybrid GPU Rendering Pipeline for Alias-Free Hard Shadows." In *Eurographics Areas Papers*, pp. 59–66, 2009.

[244] A. Hertzmann and D. Zorin. "Illustrating Smooth Surfaces." In *Proceedings of SIG-GRAPH '00, Computer Graphics Proceedings, Annual Conference Series*, pp. 517–526. New York: ACM, 2000.

[245] T.-D. Hoang and K.-L. Low. "Multi-Resolution Screen-Space Ambient Occlusion." *The Visual Computer* 27 (2011), 1–16.

[246] J. Hoberock and Y. Jia. "High-Quality Ambient Occlusion." In *GPU Gems* 3, pp. 257–274. Reading, MA: Addison Wesley, 2007.

[247] M. Hollander, T. Ritschel, E. Eisemann, and T. Boubekeur. "ManyLoDs: Parallel Many-View Level-of-Detail Selection for Real-Time Global Illumination." *Computer Graphics Forum (Eurographics Symposium on Rendering)* 30:4 (2011), 1233–1240.

[248] D. Horn, J. Sugerman, M. Houston, and P. Hanrahan. "Interactive k-d Tree GPU Raytracing." In *Proceedings of ACM Symposium on Interactive 3D Graphics and Games*, pp. 167–174. New York: ACM, 2007.

[249] S. Hornus, J. Hoerock, S. Lefebvre, and J. Hart. "ZP+: Correct Z-Pass Stencil Shadows." In *Proceedings of ACM Symposium on Interactive 3D Graphics and Games*, pp. 195–202. New York: ACM, 2005.

[250] J. Hourcade and A. Nicolas. "Algorithms for Anti-aliased Cast Shadows." *Computers and Graphics* 9:3 (1985), 259–265.

[251] H. Hu, A. Gooch, W. Thompson, B. Smits, J. Rieser, and P. Shirley. "Visual Cues for Imminent Object Contact in Realistic Virtual Environments." In *Proceedings of IEEE Visualization*, pp. 179–185. Washington, DC: IEEE Computer Society, 2000.

[252] Y. Hu, Y. Qi, and X. Shen. "A Real-Time Soft Shadow Rendering Algorithm by Occluder-Discretization." In *Congress on Image and Signal Processing*, pp. 751–755, 2008.

[253] W.A. Hunt and G.S. Johnson. "The Area Perspective Transform: A Homogeneous Transform for Efficient In-Volume Queries." *ACM Transactions on Graphics* 30:2 (2011), 8:1–6.

[254] T. Ikedo. "A Realtime Anti-aliased Soft-Shadow Casting Renderer." Technical Report HCIS-2002-01, Computer and Information Sciences, Hosei University, Japan, 2002.

[255] M. Ikits, J. Kniss, A. Lefohn, and C. Hansen. "Volume Rendering Techniques." In *GPU Gems: Programming Techniques, Tips and Tricks for Real-Time Graphics*, pp. 667–692. Reading, MA: Addison Wesley, 2004.

[256] Y. Im, C. Han, and L. Kim. "A Method to Generate Soft Shadows Using a Layered Depth Image and Warping." *IEEE Transactions on Visualization and Computer Graphics* 11:3 (2005), 265–272.

[257] T. Imagire, H. Johan, N. Tamura, and T. Nishita. "Anti-aliased and Real-Time Rendering of Scenes with Light Scattering Effects." *The Visual Computer* 23:9–11 (2007), 935–944.

[258] info@vterrain.org. "Terrain LOD: Runtime Regular-grid Algorithms." Available online (http://www.vterrain.org/LOD/Papers/).

[259] M. Isard, M. Shand, and A. Heirich. "Distributed Rendering of Interactive Soft Shadows." *Parallel Computing* 29:3 (2003), 311–323.

[260] J. Isidoro. "Shadow Mapping: GPU-Based Tips and Techniques." Presented at Game Developers Conference, 2006.

[261] T. Isshiki, M. Ishikawa, and H. Kunieda. "Cost-Effective Shadowing Method Using the ED-buffer on an Adaptive Light Cube." *The Visual Computer* 16:8 (2000), 453–468.

[262] K. Jacobs, J. Nahmias, C. Angus, A. Reche, C. Loscos, and A. Steed. "Automatic Generation of Consistent Shadows for Augmented Reality." In *Proceedings of Graphics Interface*, pp. 113–120. Toronto, Canada: Canadian Information Processing Society, 2005.

[263] B. Jakobsen, N. Christensen, B. Larsen, and K. Petersen. "Boundary Correct Real-Time Soft Shadows." In *Proceedings of Computer Graphics International*, pp. 232–239. Washington, DC: IEEE Computer Society, 2004.

[264] R. James. "True Volumetric Shadows." In *Graphics Programming Methods*, pp. 353–366. Hingham, MA: Charles River Media, 2003.

[265] J. Jansen and L. Bavoil. "Fourier Opacity Mapping." In *Proceedings of ACM Symposium on Interactive 3D Graphics and Games*, pp. 165–172. New York: ACM, 2010.

[266] F.W. Jansen and A.N.T. van der Zalm. "A Shadow Algorithm for CSG." *Computers and Graphics* 15:2 (1991), 237–247.

[267] E. Jeng and Z. Xiang. "Fast Soft Shadow Visualization for Deformable Moving Light Sources Using Adaptively Sampled Light Field Shadow Maps." In *Pacific Graphics (Posters)*, 2002.

[268] E. Jeng and Z. Xiang. "Forward Area Light Map Projection." In *Proceedings of Computer Graphics, Virtual Reality, Visualization and Interaction in Africa (AFRIGRAPH)*, pp. 79–86. New York: ACM, 2003.

[269] H. Jensen and N. Christensen. "Efficiently Rendering Shadows Using the Photon Map." In *Proceedings of Compugraphics*, pp. 285–291. Grasp, 1995.

[270] H. Jensen and N. Christensen. "Photon Maps in Bidirectional Monte Carlo Ray Tracing of Complex Objects." *Computers and Graphics* 19:2 (1995), 215–224.

[271] H. W. Jensen, J. Arvo, P. Dutré, A. Keller, A. Owen, M. Pharr, and P. Shirley. "Monte Carlo Ray Tracing." In *SIGGRAPH (Courses): Course 44*, 2003.

[272] H. Jensen. "Importance Driven Path Tracing Using the Photon Map." In *Eurographics Workshop on Rendering*, pp. 326–335. London: Springer-Verlag, 1995.

[273] H.W. Jensen. *Realistic Image Synthesis Using Photon Mapping*. Natick, MA: A K Peters, 2001.

[274] S. Jeschke, M. Wimmer, and W. Purgathofer. "Image-Based Representations for Accelerated Rendering of Complex Scenes." In *Eurographics State-of-the-Art Reports*, pp. 1–20. Aire-la-Ville, Switzerland: Eurographics, 2005.

[275] G. Johnson, W. Mark, and C. Burns. "The Irregular Z-buffer and Its Application to Shadow Mapping." Technical report, University of Texas at Austin, 2004.

[276] G. Johnson, W. Hunt, W. Mark, C. Burns, and S. Junkins. "Soft Irregular Shadow Mapping: Fast, High-Quality, and Robust Soft Shadows." In *Proceedings of ACM Symposium on Interactive 3D Graphics and Games*, pp. 57–66. New York: ACM, 2009.

[277] M. Jones. "An Efficient Shadow Detection Algorithm and the Direct Surface Rendering Volume Visualization Model." In *Proceedings of Eurographics UK*, pp. 237–244. Aire-la-ville, Switzerland: Eurographics, 1997.

[278] M. Joshi. "Non-Photorealistic Shadows." Technical report, University of Maryland at Baltimore County, 2004.

[279] V. Kajalin. "Screen Space Ambient Occlusion." In *ShaderX7: Advanced Rendering Techniques.* Hingham, MA: Charles River Media, 2009.

[280] J. Kajiya and B. Von Herzen. "Ray Tracing Volume Densities." *Proceedings of SIGGRAPH '84, Computer Graphics* 18:3 (1984), 165–174.

[281] J. Kajiya. "Ray Tracing Parametric Patches." *Proceedings of SIGGRAPH '82, Computer Graphics* 16:3 (1982), 245–254.

[282] J. Kajiya. "The Rendering Equation." *Proceedings of SIGGRAPH '86, Computer Graphics* 20:4 (1986), 143–150.

[283] T. Kakuta, T. Oishi, and K. Ikeuchi. "Shading and Shadowing of Architecture in Mixed Reality." In *Proceedings of International Symposium on Mixed and Augmented Reality (ISMAR)*, pp. 200–201. Washington, DC: IEEE Computer Society, 2005.

[284] T. Kakuta, L. Vinh, R. Kawakami, and T. Oishi. "Detection of Moving Objects and Cast Shadows Using a Spherical Vision Camera for Outdoor Mixed Reality." In *Proceedings of Symposium on Virtual Reality Software and Technology*, pp. 219–222. New York: ACM, 2008.

[285] D. Kalra and A. Barr. "Guaranteed Ray Intersection with Implicit Surfaces." *Proceedings of SIGGRAPH '89, Computer Graphics* 23:3 (1989), 297–306.

[286] T. Kanai. "Fragment-Based Evaluation of Non-Uniform B-Spline Surfaces on GPUs." *Computer Aided Design and Applications* 4:3 (2007), 287–294.

[287] Y. Kanamori, Z. Szego, and T. Nishita. "GPU-Based Fast Ray Casting for a Large Number of Metaballs." *Computer Graphics Forum (Proceedings of Eurographics)* 27:2 (2008), 351–360.

[288] M. Kanbara and N. Yokoya. "Geometric and Photometric Registration for Real-Time Augmented Reality." In *Proceedings of Symposium on Mixed and Augmented Reality*, pp. 279–280. Washington, DC: IEEE Computer Society, 2002.

[289] K. Kaneda, F. Kato, E. Nakamae, T. Nishita, H. Tanaka, and T. Noguchi. "Three dimensional Terrain Modeling and Display for Environment Assessment." *Proceedings of SIGGRAPH '89, Computer Graphics* 23:3 (1989), 207–214.

[290] S. Kashyap, R. Goradia, P. Chaudhuri, and S. Chandran. "Real Time Ray Tracing of Point-Based Models." In *Proceedings of ACM Symposium on Interactive 3D Graphics and Games.* New York: ACM, 2009.

[291] A. Katayama, Y. Sakagawa, and H. Tamura. "A Method of Shading and Shadowing in Image-Based Rendering." In *International Conference on Image Processing (ICIP)*, pp. 26–30. Los Alamitos, CA: IEEE, 1999.

[292] S. Katz, A. Tal, and R. Basri. "Direct Visibility of Point Sets." *ACM Transactions on Graphics (Proceedings of SIGGRAPH '07)* 26:3 (2007), 24:1–11.

[293]   A. Kaufman. "Voxels as a Computational Representation of Geometry." In *SIG-GRAPH (Courses)*, 1994.

[294]   J. Kautz, W. Heidrich, and K. Daubert. "Bump Map Shadows for OpenGL Rendering." Technical Report MPI-I-2000-4-001, Max-Planck-Institut Fur Informatik, 2000.

[295]   J. Kautz, P.-P. Sloan, and J. Snyder. "Fast, Arbitrary BRDF Shading for Low-Frequency Lighting Using Spherical Harmonics." In *Eurographics Workshop on Rendering*, pp. 291–296. London: Springer-Verlag, 2002.

[296]   J. Kautz, J. Lehtinen, and T. Aila. "Hemispherical Rasterization for Self-Shadowing of Dynamic Objects." In *Eurographics Symposium on Rendering*, pp. 179–184. Aire-la-Ville, Switzerland: Eurographics Association, 2004.

[297]   B. Keating and N. Max.   "Shadow Penumbras for Complex Objects by Depth-Dependent Filtering of Multi-Layer Depth Images." In *Eurographics Workshop on Rendering*, pp. 205–220. London: Springer-Verlag, 1999.

[298]   B. Keating. "Efficient Shadow Anti-aliasing Using an A-buffer." *Journal of Graphics, GPU, and Game Tools* 4:3 (1999), 23–33.

[299]   A. Keller. "Instant Radiosity." In *Proceedings of SIGGRAPH '97, Computer Graphics Proceedings, Annual Conference Series*, pp. 49–56. New York: ACM, 1997.

[300]   H. Ki, T. Ryu, and K. Oh.   "An Efficient Shadow Texture Algorithm for Shadow Maps." In *Proceedings of Computer Graphics International*, pp. 164–169. Washington, DC: IEEE Computer Society, 2008.

[301]   M. Kilgard. "Improving Shadows and Reflections via the Stencil Buffer." Course Notes for "Advanced OpenGL Game Development", presented at Game Developers Conference, http://www.philfreeman.com/stencil.pdf, 1999.

[302]   J. Kim and S. Kim. "Bilateral Filtered Shadow Maps." In *Proceedings of Symposium on Advances in Visual Computing*, pp. 49–58. Berlin: Springer-Verlag, 2009.

[303]   T. Kim and U. Neumann. "A Thin Shell Volume for Modeling Human Hair." In *Proceedings of Computer Animation*, pp. 104–115. Washington, DC: IEEE Computer Society, 2000.

[304]   T. Kim and U. Neumann. "Opacity Shadow Maps." In *Eurographics Workshop on Rendering*, pp. 177–182. London: Springer-Verlag, 2001.

[305]   T. Kim, K. Hong, I. Sato, and K. Ikeuchi. "Enhanced Augmented Reality with Shadows in Naturally Illuminated Environments." In *IAPR Workshop on Machine Vision Applications*, pp. 27–30, 2000.

[306]   B. Kim, K. Kim, and G. Turk. "A Shadow-Volume Algorithm for Opaque and Transparent Non-Manifold Casters." *Journal of Graphics, GPU, and Game Tools* 13:3 (2008), 1–14.

[307]   G. King and W. Newhall. "Efficient Omnidirectional Shadow Maps." In *ShaderX³: Advanced Rendering Techniques*, pp. 435–448. Hingham, MA: Charles River Media, 2004.

[308]   A. G. Kirk and O. Arikan. "Real-Time Ambient Occlusion for Dynamic Character Skins." In *Proceedings of ACM Symposium on Interactive 3D Graphics and Games*, pp. 47–52. New York: ACM, 2007.

[309] F. Kirsch and J. Doellner. "Real-Time Soft Shadows Using a Single Light Sample." *Journal of WSCG* 11:1–3 (2003), 255–262.

[310] J. Kniss, S. Premoze, C. Hansen, and D. Ebert. "Interactive Translucent Volume Rendering and Procedural Modeling." In *Proceedings of IEEE Visualization*, pp. 109–116. Washington, DC: IEEE Computer Society, 2002.

[311] H. Ko, D. Kang, and K. Yoon. "A Study on the Realtime Toon Rendering with Shadow." In *Proceedings of KCGS*, pp. 22–27. Korea Computer Graphics Society, 2000.

[312] T. Koneko. "Detailed Shape Representation with Parallax Mapping." In *Proceedings of International Conference on Artificial Reality and Telexistence (ICAT)*, pp. 205–208, 2001.

[313] W. Kong and M. Nakajima. "Visible Volume Buffer for Efficient Hair Expression and Shadow Generation." In *Proceedings of Computer Animation*, pp. 58–65. Washington, DC: IEEE Computer Society, 1999.

[314] J. Kontkanen and T. Aila. "Ambient Occlusion for Animated Characters." In *Eurographics Symposium on Rendering*, pp. 343–348. London: Springer-Verlag, 2006.

[315] J. Kontkanen and S. Laine. "Ambient Occlusion Fields." In *Proceedings of ACM Symposium on Interactive 3D Graphics and Games*, pp. 41–48. New York: ACM, 2005.

[316] M. Koster, J. Haber, and H. P. Seidel. "Real-Time Rendering of Human Hair Using Programmable Graphics Hardware." In *Proceedings of Computer Graphics International*, pp. 248–256. Washington, DC: IEEE Computer Society, 2004.

[317] S. Kozlov. "Perspective Shadow Maps: Care and Feeding." In *GPU Gems, Programming Techniques, Tips, and Tricks for Real-Time Graphics*, pp. 217–244. Reading, MA: Addison Wesley, 2004.

[318] A. Krishnamurthy, R. Khardekar, and S. McMains. "Optimized GPU Evaluation of Arbitrary Degree NURBS Curves and Surfaces." *Computer-Aided Design* 41:12 (2009), 971–980.

[319] J. Kruger and R. Westermann. "Acceleration Techniques for GPU-Based Volume Rendering." In *Proceedings of IEEE Visualization*, pp. 287–292. Washington, DC: IEEE Computer Society, 2003.

[320] J. Křivánek, S. Pattanaik, and J. Žára. "Adaptive Mesh Subdivision for Precomputed Radiance Transfer." In *Proceedings of Spring Conference on Computer Graphics*, pp. 106–111. New York: ACM, 2004.

[321] D. Lacewell, B. Burley, S. Boulos, and P. Shirley. "Raytracing Prefiltered Occlusion of Aggregrate Geometry." In *Proceedings of IEEE Symposium on Interactive Raytracing*, pp. 19–26. Washington, DC: IEEE Computer Society, 2008.

[322] P. Lacroute and M. Levoy. "Fast Volume Rendering Using a Shear-Warp Factorization of the Viewing Transformation." In *Proceedings of SIGGRAPH '94, Computer Graphics Proceedings, Annual Conference Series*, pp. 451–458. New York: ACM, 1994.

[323] P. Lacroute. "Fast Volume Rendering Using a Shear-Warp Factorization of the Viewing Transformation." PhD diss., Stanford University, 1995.

[324] E. Lafortune and Y. Willems. "Bi-directional Path Tracing." In *Proceedings of Compugraphics*, pp. 145–153, 1993.

[325] S. Laine and T. Aila. "Hierarchical Penumbra Casting." *Computer Graphics Forum (Proceedings of Eurographics)* 24:3 (2005), 313–322.

[326] S. Laine and T. Karras. "Efficient Sparse Voxel Octrees." In *Proceedings of ACM Symposium on Interactive 3D Graphics and Games*, pp. 55–63. New York: ACM, 2010.

[327] S. Laine and T. Karras. "Two Methods for Fast Ray-Cast Ambient Occlusion." *Computer Graphics Forum (Eurographics Symposium on Rendering)* 29:4 (2010), 1325–1333.

[328] S. Laine, T. Aila, U. Assarsson, J. Lehtinen, and T. Akenine-Möller. "Soft Shadow Volumes for Ray Tracing." *ACM Transactions on Graphics (Proceedings of SIGGRAPH '05)* 24:3 (2005), 1156–1165.

[329] S. Laine, H. Saransaari, J. Kontkanen, J. Lehtinen, and T. Aila. "Incremental Instant Radiosity for Real-Time Indirect Illumination." In *Eurographics Symposium on Rendering*. Aire-la-Ville, Switzerland: Eurographics Association, 2007.

[330] S. Laine. "Split-Plane Shadow Volumes." In *Proceedings of Graphics Hardware*, pp. 23–32. New York: ACM, 2005.

[331] A. Lake, C. Marshall, M. Harris, and M. Blackstein. "Stylized Rendering Techniques for Scalable Real-Time 3D Animation." In *Proceedings of Symposium on Non-Photorealistic Animation and Rendering*, pp. 13–20. New York: ACM, 2000.

[332] H. Landis. "Production-Ready Global Illumination." In *SIGGRAPH (Courses): Course 16*, 2002.

[333] R. Lansdale. "Shadow Depth Map Whitepaper (on the PolyTrans Software)." http://www.okino.com/new/toolkit/1-15.htm, 2005.

[334] A. Lauritzen and M. McCool. "Layered Variance Shadow Maps." In *Proceedings of Graphics Interface*, pp. 139–146. Toronto, Canada: Canadian Information Processing Society, 2008.

[335] A. Lauritzen, M. Salvi, and A. Lefohn. "Sample Distribution Shadow Maps." In *Proceedings of ACM Symposium on Interactive 3D Graphics and Games*, pp. 97–102. New York: ACM, 2011.

[336] A. Lauritzen. "Summed-Area Variance Shadow Maps." In *GPU Gems 3*, pp. 157–182. Reading, MA: Addison Wesley, 2007.

[337] C. Lauterbach, Q. Mo, and D. Manocha. "Fast Hard and Soft Shadow Generation on Complex Models Using Selective Ray Tracing." Technical Report TR09-004, UNC CS, 2009.

[338] O. Lawlor. "Interpolation-Friendly Soft Shadow Maps." In *International Conference on Computer Graphics and Virtual Reality*, pp. 111–117. Providence, RI: CSREA, 2006.

[339] A. LeBlanc, R. Turner, and D. Thalmann. "Rendering Hair Using Pixel Blending and Shadow buffers." *Journal of Visualization and Computer Animation* 2:3 (1991), 92–97.

[340] M. Lee, R. Redner, and S. Uselton. "Statistically Optimized Sampling for Distributed Ray Tracing." *Proceedings of SIGGRAPH '85, Computer Graphics* 19:3 (1985), 61–68.

[341] A. Lefohn, S. Sengupta, and J. Kniss. "Dynamic Adaptive Shadow Maps on Graphics Hardware." In *SIGGRAPH (Sketches)*. New York: ACM, 2005.

[342] A. Lefohn, S. Sengupta, and J. Owens. "Resolution-Matched Shadow Maps." *ACM Transactions on Graphics* 26:4 (2007), 1–23.

[343] J. Lehtinen, S. Laine, and T. Aila. "An Improved Physically-Based Soft Shadow Volume Algorithm." *Computer Graphics Forum (Proceedings of Eurographics)* 25:3 (2006), 303–312.

[344] J. Lengyel and J. Snyder. "Rendering with Coherent Layers." In *Proceedings of SIG-GRAPH '97, Computer Graphics Proceedings, Annual Conference Series*, pp. 233–242. New York: ACM, 1997.

[345] E. Lengyel. "The Mechanics of Robust Stencil Shadows." *Gamasutra*, (October, 2002): http://www.gamasutra.com/view/feature/2942/the_mechanics_of_robust_stencil_.php.

[346] E. Lengyel. "Advanced Stencil Shadow and Penumbral Wedge Rendering." Presented at Game Developers Conference, 2005.

[347] M. Levoy and T. Whitted. "The Use of Points as a Display Primitive." Technical Report TR 85-022, Department of Computer Science, University of North Carolina at Chapel Hill, 1985.

[348] M. Levoy. "Display of Surfaces from Volume Data." *IEEE Computer Graphics and Applications* 8:3 (1988), 29–37.

[349] D. Lichtenberger. "Volumetric Shadows." Technical report, Institute of Computer Graphics and Algorithms, Technical University of Vienna, 2003.

[350] P. Lindstrom, D. Koller, W. Ribarsky, L. Hodges, N. Faust, and G. Turner. "Real-Time, Continous Level of Detail Rendering of Height Fields." In *Proceedings of SIG-GRAPH '96, Computer Graphics Proceedings, Annual Conference Series*, pp. 109–118. New York: ACM, 1996.

[351] L. Linsen, K. Muller, and P. Rosenthal. "Splat-Based Ray Tracing of Point Clouds." *Journal of WSCG* 15:1–3 (2007), 51–58.

[352] D. Lischinski and A. Rappoport. "Image-Based Rendering for Non-Diffuse Synthetic Scenes." In *Eurographics Workshop on Rendering*, pp. 301–314. London: Springer-Verlag, 1998.

[353] D. Lischinski, F. Tampieri, and D. Greenberg. "Discontinuity Meshing for Accurate Radiosity." *IEEE Computer Graphics and Applications* 12:6 (1992), 25–39.

[354] N. Liu and M. Pang. "An Introduction of Shadow Mapping Rendering Algorithms in Realistic Computer Graphics." Presented at *International Workshop on Information Security and Application*, 2009.

[355] N. Liu and M. Pang. "A Survey of Shadow Rendering Algorithms: Projection Shadows and Shadow Volumes." In *International Workshop on Computer Science and Engineering (IWCSE)*, pp. 488–492. Washington, DC: IEEE Computer Society, 2009.

[356] M. Liu and E. Wu. "A Real-Time Algorithm for 3-D Generation for Point Light Sources." *Journal of Software* 11:6 (2000), 785–790.

[357] D.B. Lloyd, J. Wendt, N. Govindaraju, and D. Manocha. "CC Shadow Volumes." In *Eurographics Symposium on Rendering*, pp. 197–206. Aire-la-Ville, Switzerland: Eurographics Association, 2004.

[358] D. Lloyd, D. Tuft, S. Yoon, and D. Manocha. "Warping and Partioning for Low Error Shadow Maps." In *Eurographics Symposium on Rendering*, pp. 215–226. Aire-la-Ville, Switzerland: Eurographics Association, 2006.

[359] B. Lloyd, N.K. Govindara, S.E. Molnar, and D. Manocha. "Practical Logarithmic Rasterization for Low-error Shadow Maps." In *Proceedings of Graphics Hardware*, pp. 17–24. New York: ACM, 2007.

[360] D. Lloyd, N. Govindaraju, C. Quammen, S. Molnar, and D. Manocha. "Logarithmic Perspective Shadow Maps." *ACM Transactions on Graphics* 27:4 (2008), 106:1–32.

[361] J. Lluch, E. Camahort, and R. Vivó. "An Image-Based Multiresolution Model for Interactive Foliage Rendering." *Journal of WSCG* 12:1–3 (2004), 507–514.

[362] T. Lokovic and E. Veach. "Deep Shadow Maps." In *Proceedings of SIGGRAPH '00, Computer Graphics Proceedings, Annual Conference Series*, pp. 385–392. New York: ACM, 2000.

[363] C. Loop and J. Blinn. "Real-Time GPU Rendering of Piecewise Algebraic Surfaces." *ACM Transactions on Graphics (Proceedings of SIGGRAPH '06)* 25:3 (2006), 664–670.

[364] B.J. Loos and P.-P. Sloan. "Volumetric Obscurance." In *Proceedings of ACM Symposium on Interactive 3D Graphics and Games*, pp. 151–156. New York: ACM, 2010.

[365] P.F. Lopes. "Topics in Animation: The Pinscreen in the Era of the Digital Image." http://www.writer2001.com/lopes.htm, 2001.

[366] P. Lord and B. Sibley. *Creating 3D Animation: The Aardman Book of Filmmaking.* New York: Harry N. Abrams Inc., 1998.

[367] W. Lorensen and H. Cline. "Marching Cubes: A High Resolution 3D Surface Reconstruction Algorithm." *Proceedings of SIGGRAPH '87, Computer Graphics* 21:4 (1987), 163–169.

[368] C. Loscos and G. Drettakis. "Interactive High-Quality Soft Shadows in Scenes with Moving Objects." *Computer Graphics Forum (Proceedings of Eurographics)* 16:3 (1997), 219–230.

[369] C. Loscos, M. Frasson, G. Drettakis, and B. Walter. "Interactive Virtual Relighting and Remodelling of Real Scenes." In *Eurographics Workshop on Rendering*, pp. 329–340. London: Springer-Verlag, 1999.

[370] C. Loscos, F. Tecchia, and Y. Chrysanthou. "Real-Time Shadows for Animated Crowds in Virtual Cities." In *Proceedings of the ACM Symposium on Virtual Reality, Software and Technology*, pp. 85–92. New York: ACM, 2001.

[371] B. Lu, T. Kakuta, R. Kawakami, and T. Oishi. "Foreground and Shadow Occlusion Handling for Outdoor Augmented Reality." In *Proceedings of International Symposium on Mixed and Augmented Reality*, pp. 109–118. Los Alamitos, CA: IEEE, 2010.

[372] A. Lukaszewski. "Exploiting Coherence of Shadow Rays." In *Proceedings of the International Conference on Computer Graphics, Virtual Reality and Visualization*, pp. 147–150. New York: ACM, 2001.

[373] W. Lv, X. Liu, and E. Wu. "Practical Hybrid Pre-Filtering Shadow Maps." In *ACM International Conference on Virtual-Reality Continuum and Its Applications in Industry (VRCAI)*, pp. 52:1–2. New York: ACM, 2008.

[374] P. Maciel and P. Shirley. "Visual Navigation of Large Environments Using Textured Clusters." In *Proceedings of ACM Symposium on Interactive 3D Graphics*, pp. 95–102. New York: ACM, 1995.

[375] C. Madison, W. Thompson, D. Kersten, P. Shirley, and B. Smits. "Use of Interreflection and Shadow for Surface Contact." *Perception and Psychologist* 63 (2001), 187–194.

[376] C. Madsen and R. Laursen. "A Scalable GPU-Based Approach to Shading and Shadowing for Photorealistic Real-Time Augmented Reality." In *Computer Graphics Theory and Applications*, pp. 252–261, 2007. Available online (http://www.informatik.uni-trier-.de/~ley/db/conf/grapp/grapp2001-1.html).

[377] C. Madsen, M. Sorensen, and M. Vittrup. "The Importance of Shadows in Augmented Reality." Presented at *International Workshop on Presence*, 2003. Available online (http://vbn.aau.dk/files/16358238/dankom08probelessAR.pdf).

[378] C. Madsen. "Estimating Radiances of the Sun and the Sky from a Single Image Containing Shadows." In *Danish Conference on Pattern Recognition and Image Analysis*, 2008.

[379] M. Malmer, F. Malmer, U. Assarsson, and N. Holzschuch. "Fast Precomputed Ambient Occlusion for Proximity Shadows." *Journal of Graphics, GPU, and Game Tools* 12:2 (2007), 59–71.

[380] P. Mamassiam, D. Knill, and D. Kersten. "The Perception of Cast Shadows." *Trends in Cognitive Science* 2:8 (1998), 288–295.

[381] M. Marghidanu. "Fast Computation of Terrain Shadow Maps." http://www.gamedev.net/reference/articles/article1817.asp, 2002.

[382] L. Markosian, M. Kowalski, S. Trychin, L. Bourdev, D. Goldstein, and J. Hughes. "Real-Time Nonphotorealistic Rendering." In *Proceedings of SIGGRAPH '97, Computer Graphics Proceedings, Annual Conference Series*. New York: ACM, 1997.

[383] L. Markosian. "Non-Photorealistic Rendering." COS 426 Guest Lecture, Princeton University, 2003.

[384] J. Marlon. *Focus on Photon Mapping*. Tulsa, OK: Premier Press, 2003.

[385] G. Marmitt, H. Friedrich, and P. Slusallek. "Interactive Volume Rendering with Ray Tracing." In *Eurographics State-of-the-Art Report*. Aire-la-Ville, Switzerland: Eurographics, 2006.

[386] T. Martin and T. Tan. "Anti-aliasing and Continuity with Trapezoidal Shadow Maps." In *Eurographics Symposium on Rendering*, pp. 153–160. Aire-la-Ville, Switzerland: Eurographics Association, 2004.

[387] T. Martin. "Perspective Shadow Mapping: Implementation, Results, Problems." Diploma thesis, National University of Singapore, 2003.

[388] T. Martinec. "Real-Time Non-Photorealistic Shadow Rendering." Master's thesis, Abo Akademi University, 2007.

[389] N. Max and K. Ohsaki. "Rendering Trees from Precomputed Z-buffer Views." In *Eurographics Workshop on Rendering*, pp. 45–54. London: Springer-Verlag, 1995.

[390] N. Max. "Atmospheric Illumination and Shadows." *Proceedings of SIGGRAPH '86, Computer Graphics* 20:4 (1986), 117–124.

[391] N. Max. "Horizon Mapping: Shadows for Bump Mapped Surfaces." *The Visual Computer* 4:2 (1988), 109–117.

[392] J.R. Maxwell, J. Beard, S. Weiners, D. Ladd, and J. Ladd. "Bidirectional Reflectance Model Validation and Utilization." Technical Report AFAL-TR-73-303, Environmental Research Institute of Michigan (ERIM), 1973.

[393] M. McCool. "Shadow Volume Reconstruction from Depth Maps." *ACM Transactions on Graphics* 19:1 (2000), 1–26.

[394] M. McGuire and E. Enderton. "Colored Stochastic Shadow Maps." In *Proceedings of ACM Symposium on Interactive 3D Graphics and Games*, pp. 89–96. New York: ACM, 2011.

[395] M. McGuire and A. Fein. "Real-Time Rendering of Cartoon Smoke and Clouds." In *Proceedings of Symposium on Non-Photorealistic Animation and Rendering*, pp. 21–26. New York: ACM, 2006.

[396] M. McGuire and M. McGuire. "Steep Parallax Mapping." In *ACM Symposium on Interactive 3D Graphics and Games (Posters)*, 2005.

[397] M. McGuire, J. Hughes, K. Egan, M. Kilgard, and C. Everitt. "Fast, Practical and Robust Shadows." Technical report, NVIDIA, 2003.

[398] M. McGuire. "Efficient Shadow Volume Rendering." In *GPU Gems, Programming Techniques, Tips and Tricks for Real-Time Graphics*, pp. 137–166. Reading, MA: Addison Wesley, 2004.

[399] M. McGuire. "Observations on Silhouette Sizes." *Journal of Graphics, GPU, and Game Tools* 9:1 (2004), 1–12.

[400] M. McGuire. "Single-Pass Shadow Volumes for Arbitrary Meshes." In *SIGGRAPH (Posters)*, p. 177, 2007.

[401] M. McGuire. "Ambient Occlusion Volumes." In *Proceedings of High Performance Graphics*, pp. 47–56. New York: ACM, 2010.

[402] D. Meaney and C. O'Sullivan. "Heuristical Real-Time Shadows." Technical Report TCD-DS-1999-19, Trinity College Dublin, 1999.

[403] A. Mendez-Feliu and M. Sbert. "From Obscurances to Ambient Occlusion: A Survey." *The Visual Computer* 25:2 (2009), 181–196.

[404] A. Mendez-Feliu, M. Sbert, and J. Cata. "Real-Time Obscurances with Color Bleeding." In *Proceedings of Spring Conference on Computer Graphics*, pp. 171–176. New York: ACM, 2003.

[405] T. Mertens, J. Kautz, P. Bekaert, and F. van Reeth. "A Self-Shadow Algorithm for Dynamic Hair Using Clustered Densities." In *Eurographics Symposium on Rendering*, pp. 173–178. Aire-la-Ville, Switzerland: Eurographics Association, 2004.

[406] A. Meyer, F. Neyret, and P. Poulin. "Interactive Rendering of Trees with Shading and Shadows." In *Eurographics Workshop on Rendering*, pp. 183–196. London: Springer-Verlag, 2001.

[407] U. Meyer. "Hemi-cube Ray Tracing: A Method for Generating Soft Shadows." 9 (1990), 365–376.

[408] M. Mikkelsea. "Separating-Planes Perspective Shadow Mapping." *Journal of Graphics, GPU, and Game Tools* 12:3 (2007), 43–54.

[409] G. Miller. "The Definition and Rendering of Terrain Maps." *Proceedings of SIG-GRAPH '86, Computer Graphics* 20:4 (1986), 39–48.

[410] G. Miller. "Efficient Algorithms for Local and Global Accessibility Shading." In *Proceedings of SIGGRAPH '94, Computer Graphics Proceedings, Annual Conference Series*, pp. 319–326. New York: ACM, 1994.

[411] D. Mitchell. "Robust Ray Tracing Intersection with Interval Arithmetic." In *Proceedings of Graphics Interface*, pp. 68–74. Toronto, Canada: Canadian Information Processing Society, 1990.

[412] J. Mitchell. "Light Shafts: Rendering Shadows in Participating Media." Presented at Game Developers Conference, 2004.

[413] K. Mitchell. "Volumetric Light Scattering as a Post-process." In *GPU Gems 3*, pp. 275–285. Reading, MA: Addison Wesley, 2007.

[414] N. Mitra and M. Pauly. "Shadow Art." *ACM Transactions on Graphics (Proceedings of SIGGRAPH Asia '09)* 28:5 (2009), 156:1–7.

[415] M. Mittring. "Finding Next Gen—CryEngine 2." In *SIGGRAPH (Courses): Advanced Real-Time Rendering in 3D Graphics and Games*, 2007.

[416] Q. Mo, V. Popescu, and C. Wyman. "The Soft Shadow Occlusion Camera." In *Proceedings of Pacific Graphics*, pp. 189–198. Washington, DC: IEEE Computer Society, 2007.

[417] M. MohammadBagher, J. Kautz, N. Holzschuch, and C. Soler. "Screen-Space Percentage-Closer Soft Shadows." In *SIGGRAPH (Posters)*, 2010.

[418] F. Mora and L. Aveneau. "Fast and Exact Direct Illumination." In *Proceedings of Computer Graphics International*, pp. 191–197. Washington, DC: IEEE Computer Society, 2005.

[419] K. Mueller and R. Crawfis. "Eliminating Popping Artifacts in Sheet Buffer-Based Splatting." In *Proceedings of Visualization*, pp. 239–246. Los Alamitos, CA: IEEE Computer Society Press, 1998.

[420] K. Musgrave. "Grid Tracing: Fast Ray Tracing for Height Fields." Technical Report YALEU/DCS/RR-639, Yale University, 1988.

[421] G. Nagy. "Real-Time Shadows on Complex Objects." In *Game Programming Gems*, pp. 567–580. Hingham, MA: Charles River Media, 2000.

[422] G. Nakano, I. Kitahara, and Y. Ohta. "Generating Perceptually-Correct Shadows for Mixed Reality." In *Proceedings of International Symposium on Mixed and Augmented Reality*, pp. 173–174. Los Alamitos, CA: IEEE, 2008.

[423] R. Ng, R. Ramamoorthi, and P. Hanrahan. "All-Frequency Shadows Using Non-Linear Wavelet Lighting Approximation." *ACM Transactions on Graphics (Proceedings of SIGGRAPH '03)* 22:3 (2003), 376–381.

[424] H. Nguyen and W. Donnelly. "Hair Animation and Rendering in the Nalu Demo." In *GPU Gems 2*, pp. 361–380. Reading, MA: Addison Wesley, 2005.

[425] K. Nguyen, H. Jang, and J. Han. "Layered Occlusion Map for Soft Shadow Generation." *The Visual Computer* 26:12 (2010), 1497–1512.

[426]   H. Nguyen. "Casting Shadows on Volume." *Game Developer Magazine* 6:3 (1999), 44–53.

[427]   R. Ni, M. Braunstein, and G. Andersen. "Interaction of Optical Contact, Shadows and Motion in Determining Scene Layout." *Journal of Vision* 4:8 (2004), 615.

[428]   T. Nishita and E. Nakamae. "An Algorithm for Half-Tone Representation of Three-Dimensional Objects." *Information Processing in Japan* 14 (1974), 93–99.

[429]   T. Nishita and E. Nakamae. "Continuous Tone Representation of Three-Dimensional Objects Illuminated by Sky Light." *Proceedings of SIGGRAPH '86, Computer Graphics* 20:4 (1986), 125–132.

[430]   T. Nishita, I. Okamura, and E. Nakamae. "Shading Models for Point and Linear Sources." *ACM Transactions on Graphics* 4:2 (1985), 124–146.

[431]   T. Nishita, Y. Miyawaki, and E. Nakamae. "A Shading Model for Atmospheric Scattering Considering Luminous Intensity Distribution of Light Sources." *Proceedings of SIGGRAPH '87, Computer Graphics* 21:4 (1987), 303–310.

[432]   T. Nishita, T. Sederberg, and M. Kakimoto. "Ray Tracing Trimmed Rational Surface Patches." *Proceedings of SIGGRAPH '90, Computer Graphics* 24:4 (1990), 337–345.

[433]   Z. Noh and M. Sunar. "A Review of Shadow Techniques in Augmented Reality." In *International Conference on Machine Vision*, pp. 320–324. Washington, DC: IEEE Computer Society, 2009.

[434]   Z. Noh and M. Sunar. "Soft Shadow Rendering Based on Real Light Source Estimation in Augmented Reality." In *Advances in Multimedia: An International Journal (AMIJ)* 1:2 (2010), 26–36.

[435]   T. Noma and K. Sumi. "Shadows on Bump-Mapped Surfaces in Ray Tracing." *The Visual Computer* 10:6 (1994), 330–336.

[436]   M. Nulkar and K. Mueller. "Splatting with Shadows." In *Proceedings of Volume Graphics*, pp. 35–50, 2001.

[437]   K. Oh and S. Park. "Realtime Hybrid Shadow Algorithm Using Shadow Texture and Shadow Map." In *ICCSA*, pp. 972–980. Berlin, Heidelberg: Springer-Verlag, 2007.

[438]   K. Oh, B. Shin, and Y. Shin. "Linear Time Shadow Texture Generation Algorithm." Presented at *Game Technology Conference (GTEC)*, 2001.

[439]   K. Onoue, N. Max, and T. Nishita. "Real-Time Rendering of Bumpmap Shadows Taking Account of Surface Curvature." In *International Conference on Cyberworlds*, pp. 312–318. Washington, DC: IEEE Computer Society, 2004.

[440]   M. Oren and S.K. Nayar. "Generalization of Lambert's Reflectance Model." In *Proceedings of SIGGRAPH '94, Computer Graphics Proceedings, Annual Conference Series*, pp. 239–246. New York: ACM, 1994.

[441]   B. Osman, B. Bukowski, and C. McEvoy. "Practical Implementation of Dual Paraboloid Shadow Maps." In *Proceedings of ACM SIGGRAPH Symposium on Videogames*, pp. 397–408. New York: ACM, 2006.

[442]   V. Ostromoukhov, C. Donohue, and P.-M. Jodoin. "Fast Hierarchical Importance Sampling with Blue Noise Properties." *ACM Transactions on Graphics (Proceedings of SIGGRAPH '04)* 23:3 (2004), 488–495.

[443] J. Ou and F. Pellacini. "LightSlice: Matrix Slice Sampling for the Many-Lights Problem." *ACM Transactions on Graphics (Proceedings of SIGGRAPH Asia '11)* 30:6 (2011), 26:1–10.

[444] M. Ouellette and E. Fiume. "Approximating the Location of Integrand Discontinuities for Penumbral Illumination with Area Light Sources." In *Eurographics Workshop on Rendering*, pp. 213–224. London: Springer-Verlag, 1999.

[445] M. Ouellette and E. Fiume. "Approximating the Location of Integrand Discontinuities for Penumbral Illumination with Linear Light Sources." In *Proceedings of Graphics Interface*, pp. 66–75. Toronto, Canada: Canadian Information Processing Society, 1999.

[446] M. Ouellette and E. Fiume. "On Numerical Solutions to One-Dimensional Integration Problems with Applications to Linear Light Sources." *ACM Transactions on Graphics* 20:4 (2001), 232–279.

[447] R. Overbeck, R. Ramamoorthi, and W. Mark. "A Real-Time Beam Tracer with Application to Exact Soft Shadows." In *Eurographics Symposium on Rendering*, pp. 85–98. Aire-la-Ville, Switzerland: Eurographics Association, 2007.

[448] H. Pabst, J. Springer, A. Schollmeyer, R. Lenhardt, C. Lessig, and B. Froehlich. "Ray Casting Trimmed NURBS Surfaces on the GPU." In *Proceedings of IEEE Symposium on Interactive Ray Tracing*, pp. 151–160. Washington, DC: IEEE Computer Society, 2006.

[449] P. Pacotto. "Art Gallery, St.Paul de Vence, France." http://www.pacotto.com/sculptures.htm.

[450] C. Pagot, J. Comba, and M. Neto. "Multiple-Depth Shadow Maps." In *Computer Graphics and Image Processing*, pp. 308–315. Washington, DC: IEEE Computer Society, 2004.

[451] M. Pan, R. Wang, W. Chen, K. Zhou, and H. Bao. "Fast, Sub-pixel Antialiased Shadow Maps." *Computer Graphics Forum* 28:7 (2009), 1927–1934.

[452] S. Parker, P. Shirley, and B. Smits. "Single Sample Soft Shadows." Technical Report TR UUCS-98-019, Computer Science Department, University of Utah, 1998.

[453] A. Pearce and D. Jevans. "Exploiting Shadow Coherence in Ray Tracing." In *Proceedings of Graphics Interface*, pp. 109–116. Toronto, Canada: Canadian Information Processing Society, 1991.

[454] A. Pearce. "Shadow Attenuation for Ray Tracing Transparent Objects." In *Graphics Gems*, pp. 397–399. New York: Academic Press, 1990.

[455] F. Pellacini, P. Tole, and D. Greenberg. "A User Interface for Interactive Cinematic Shadow Design." *ACM Transactions on Graphics (Proceedings of SIGGRAPH '02)* 21:3 (2002), 563–566.

[456] F. Pellacini, F. Battaglia, R. Morley, and A. Finkelstein. "Lighting with Paint." *ACM Transactions on Graphics* 26:2 (2007), 9:1–14.

[457] Carin Perron. "Topics in Animation: Alexander Alexeieff and Claire Parker." http://www.writer2001.com/analexei.htm, 2001.

[458] L. Petrović, B. Fujito, L. Williams, and A. Finkelstein. "Shadows for Cel Animation." In *Proceedings of SIGGRAPH '00, Computer Graphics Proceedings, Annual Conference Series*, pp. 511–516. New York: ACM, 2000.

[459] H. Pfister, M. Zwicker, J. van Baar, and M. Gross. "Surfels: Surface Elements as Rendering Primitives." In *Proceedings of SIGGRAPH '00, Computer Graphics Proceedings, Annual Conference Series*, pp. 335–342. New York: ACM, 2000.

[460] M. Pharr and S. Green. "Ambient Occlusion." In *GPU Gems, Ch. 17*. Reading, MA: Addison Wesley, 2004.

[461] M. Pharr, C. Kolb, R. Gershbein, and P. Hanrahan. "Rendering Complex Scenes with Memory-Coherent Ray Tracing." In *Proceedings of SIGGRAPH '97, Computer Graphics Proceedings, Annual Conference Series*, pp. 101–108. New York: ACM, 1997.

[462] B.T. Phong. "Illumination for Computer Generated Pictures." *Communications of the ACM* 18:6 (1975), 311–317.

[463] K. Picott. "Extensions of the Linear and Area Lighting Models." *IEEE Computer Graphics and Applications* 12:2 (1992), 31–38.

[464] S. Popov, J. Gunther, H. P. Seidel, and P. Slusallek. "Stackless kd-tree Traversal for High Performance GPU Ray Tracing." *Computer Graphics Forum* 26:3 (2007), 415–424.

[465] P. Poulin and J. Amanatides. "Shading and Shadowing with Linear Light Sources." In *Proceedings of Eurographics*, pp. 377–386. Aire-la-ville Switzerland: Eurographics, 1990.

[466] P. Poulin and A. Fournier. "A Model for Anisotropic Reflection." *Proceedings of SIGGRAPH '90, Computer Graphics* 24:4 (1990), 273–282.

[467] P. Poulin and A. Fournier. "Lights from Highlights and Shadows." In *Proceedings of ACM Symposium on Interactive 3D Graphics*, pp. 31–38. New York: ACM, 1992.

[468] P. Poulin, K. Ratib, and M. Jacques. "Sketching Shadows and Highlights to Position Lights." In *Proceedings of Computer Graphics International*, pp. 56–63. Washington, DC: IEEE Computer Society, 1997.

[469] T. Purcell, I. Buck, W. Mark, and P. Hanrahan. "Ray Tracing on Programmable Graphics Hardware." *ACM Transactions on Graphics (Proceedings of SIGGRAPH '02)* 21:3 (2002), 703–712.

[470] X. Qin, E. Nakamae, K. Tadamura, and Y. Nagai. "Fast Photo-Realistic Rendering of Trees in Daylight." *Computer Graphics Forum (Proceedings of Eurographics)* 22:3 (2003), 243–252.

[471] H. Qu, F. Qiu, N. Zhang, A. Kaufman, and M. Wan. "Ray Tracing Height Fields." In *Proceedings of Computer Graphics International*, pp. 202–207. Washington, DC: IEEE Computer Society, 2003.

[472] P. Rademacher, P. Lengyel, J. Cutrell, and T. Whitted. "Measuring the Perception of Visual Realism in Images." In *Eurographics Workshop on Rendering*, pp. 235–248. London: Springer-Verlag, 2001.

[473] R. Ramamoorthi, D. Mahajan, and P. Belhumeur. "A First-Order Analysis of Lighting, Shading, and Shadows." *ACM Transactions on Graphics* 26:1 (2007), 1.

[474] R. Ramamoorthi. "Precomputation-Based Rendering." *Foundations and Trends in Computer Graphics and Vision* 3:4 (2009), 281–369.

[475] W. Reeves and R. Blau. "Approximate and Probabilistic Algorithms for Shading and Rendering Structured Particle Systems." *Proceedings of SIGGRAPH '85, Computer Graphics* 19:3 (1985), 313–322.

[476] W. Reeves, D. Salesin, and R. Cook. "Rendering Anti-aliased Shadows with Depth Maps." *Proceedings of SIGGRAPH '87, Computer Graphics* 21:4 (1987), 283–291.

[477] C. Reinbothe, T. Boubekeur, and M. Alexa. "Hybrid Ambient Occlusion." In *Eurographics (Short Papers).* Aire-la-Ville, Switzerland: Eurographics Association, 2009.

[478] Z. Ren, R. Wang, J. Snyder, K. Zhou, X. Liu, B. Sun, P.-P. Sloan, H. Bao, Q. Peng, and B. Guo. "Real-Time Soft Shadows in Dynamic Scenes Using Spherical Harmonic Exponentiation." *ACM Transactions on Graphics (Proceedings of SIGGRAPH '06)* 25:3 (2006), 977–986.

[479] T. Ritschel, R. Grosch, J. Kautz, and S. Muller. "Interactive Illumination with Coherent Shadow Maps." In *Eurographics Symposium on Rendering,* pp. 61–72. Aire-la-Ville, Switzerland: Eurographics Association, 2007.

[480] T. Ritschel, T. Grosch, J. Kautz, and H. P. Seidel. "Interactive Global Illumination Based on Coherent Surface Shadow Maps." In *Proceedings of Graphics Interface,* pp. 185–192. Toronto, Canada: Canadian Information Processing Society, 2008.

[481] T. Ritschel, T. Grosch, M. Kim, H. P. Seidel, C. Dachsbacher, and J. Kautz. "Imperfect Shadow Maps for Efficient Computation of Indirect Illumination." *ACM Transactions on Graphics (Proceedings of SIGGRAPH Asia '08)* 27:5 (2008), 1–8.

[482] T. Ritschel, T. Grosch, and H. P. Seidel. "Approximating Dynamic Global Illumination in Image Space." In *Proceedings of ACM Symposium on Interactive 3D Graphics and Games,* pp. 75–82. New York: ACM, 2009.

[483] F. Ritter and T. Strohotte. "Using Semi-Transparent Shadow Volumes to Facilitate Direct Manipulation in Interactive Environments." Submitted for publication: http://isgwww.cs.uni-magdeburg.de/~fritter/bdy-ssv.html, 2003.

[484] F. Ritter, H. Sonnet, K. Hartmann, and T. Strothotte. "Illustrative Shadows: Integrating 3D and 2D Information Displays." In *Proceedings of Intelligent User Interfaces,* pp. 166–173, 2003.

[485] P. Robertson. "Spatial Transformations for Rapid Scan-line Surface Shadowing." *IEEE Computer Graphics and Applications* 9:2 (1989), 30–38.

[486] A. Robison and P. Shirley. "Image Space Gathering." In *Proceedings of High Performance Graphics,* pp. 91–98. New York: ACM, 2009.

[487] S. Roettger, A. Irion, and T. Ertl. "Shadow Volumes Revisited." *Journal of WSCG* 10:1–3 (2002), 373–393.

[488] S. Roettger, S. Guthe, D. Weiskopf, and T. Ertl. "Smart Hardware-Accelerated Volume Rendering." In *Proceedings of Symposium on Visualization (VisSym),* pp. 231–238. Aire-la-Ville, Switzerland: Eurographics Association, 2003.

[489] G. Rong and T. Tan. "Utilizing Jump Flooding in Image-Based Soft Shadows." In *Proceedings of ACM Symposium on Virtual Reality Software and Technology,* pp. 173–180. New York: ACM, 2006.

[490] T. Ropinski, J. Kasten, and K. Hinrichs. "Efficient Shadows for GPU-Based Volume Raycasting." *Journal of WSCG* 16:1–3 (2008), 17–24.

[491] T. Ropinski, C. Doring, and C. Rezk-Salama. "Interactive Volumetric Lighting Simulating Scattering and Shadowing." In *Proceedings of Asia Pacific Symposium on Information Visualization,* pp. 169–176. Los Alamitos, CA: IEEE, 2010.

[492] H. Rushmeier and K. Torrance. "The Zonal Method for Calculating Light Intensities in the Presence of a Participating Medium." *Proceedings of SIGGRAPH '87, Computer Graphics* 21:4 (1987), 293–302.

[493] G. Ryder and A. Day. "High Quality Shadows for Real-Time Crowds." In *Eurographics (Short Papers),* pp. 37–41. Aire-la-Ville, Switzerland: Eurographics Association, 2006.

[494] D. Salesin and J. Stolfi. "The ZZ-buffer: A Simple and Efficient Rendering Algorithm with Reliable Antialiasing." Presented at *PIXIM '89,* 1989.

[495] M. Salvi, K. Vidimce, A. Lauritzen, and A. Lefohn. "Adaptive Volumetric Shadow Maps." *Computer Graphics Forum (Eurographics Symposium on Rendering)* 29:4 (2010), 1289–1296.

[496] M. Salvi. "Rendering Filtered Shadows with Exponential Shadow Maps." In *ShaderX^6: Advanced Rendering Techniques,* pp. 257–274. Hingham, MA: Charles River Media, 2008.

[497] I. Sato, Y. Sato, and K. Ikeuchi. "Illumination Distribution from Brightness in Shadows: Adaptive Estimation of Illumination Distribution with Unknown Reflectance Properties in Shadow Regions." In *International Conference on Computer Vision,* pp. 875–882. Los Alamitos, CA: IEEE, 1999.

[498] I. Sato, Y. Sato, and K. Ikeuchi. "Illumination from Shadows." *IEEE Transactions on Pattern Analysis and Machine Intelligence* 25:3 (2003), 290–300.

[499] M. Sattler, R. Sarlette, T. Mucken, and R. Klein. "Exploitation of Human Shadow Perception for Fast Shadow Rendering." In *Proceedings of Symposium on Applied Perception in Graphics and Visualization,* pp. 131–134. New York: ACM, 2004.

[500] F. Sauer, O. Masclef, Y. Robert, and P. Deltour. "Shadow Effects in Outcast." Presented at Game Developers Conference, 1999. Available online (http://www.appeal. be/products/page1/Outcast_GDC/outcast_gdc_10.htm).

[501] R. Sayeed and T. Howard. "State-of-the-Art Non-Photorealistic Rendering (NPR) Techniques." In *Theory and Practice of Computer Graphics,* edited by M. McDerby and L. Lever. pp. 1–10, 2006.

[502] G. Schaufler and H. Jensen. "Ray Tracing Point Sampled Geometry." In *Eurographics Workshop on Rendering,* pp. 319–328. London: Springer-Verlag, 2000.

[503] G. Schaufler, J. Dorsey, X. Décoret, and F. Sillion. "Conservative Volumetric Visibility with Occluder Fusion." In *Proceedings of SIGGRAPH '00, Computer Graphics Proceedings, Annual Conference Series,* pp. 229–238. New York: ACM, 2000.

[504] A. Scheel, M. Stamminger, and H. P. Seidel. "Thrifty Final Gather for Radiosity." In *Eurographics Workshop on Rendering,* pp. 1–12. London: Springer-Verlag, 2001.

[505] D. Scherzer, S. Jeschke, and M. Wimmer. "Pixel-Correct Shadow Maps with Temporal Reprojection and Shadow Test Confidence." In *Eurographics Symposium on Rendering,* pp. 45–50. Aire-la-Ville, Switzerland: Eurographics Association, 2007.

[506] D. Scherzer, M. Schwarzler, O. Mattausch, and M. Wimmer. "Real-Time Soft Shadows Using Temporal Coherence." In *ISVC*, pp. 13–24. Berlin: Springer-Verlag, 2009.

[507] D. Scherzer, M. Wimmer, and W. Purgathofer. "A Survey of Real-Time Hard Shadow Mapping Methods." In *Eurographics State-of-the-Art Reports*, pp. 21–36. Aire-la-Ville, Switzerland: Eurographics, 2010.

[508] D. Scherzer. "Robust Shadow Maps for Large Environments." In *Proceedings of Central European Seminar on Computer Graphics*, pp. 15–22, 2005. Available online (http://www.cescg.org/proceedings.html).

[509] C. Schüler. "Eliminating Surface Acne with Gradient Shadow Mapping." In *ShaderX⁴: Advanced Rendering Techniques*, pp. 289–297. Hingham, MA: Charles River Media, 2005.

[510] C. Schüler. "Multisampling Extension for Gradient Shadow Maps." In *ShaderX⁵: Advanced Rendering Techniques*, pp. 207–218. Hingham, MA: Charles River Media, 2006.

[511] M. Schwarz and M. Stamminger. "Bitmask Soft Shadows." *Computer Graphics Forum (Proceedings of Eurographics)* 26:3 (2007), 515–524.

[512] M. Schwarz and M. Stamminger. "Microquad Soft Shadow Mapping Revisited." In *Eurographics (Short Papers)*, pp. 295–298. Aire-la-Ville, Switzerland: Eurographics Association, 2008.

[513] M. Schwarz and M. Stamminger. "Quality Scalability of Soft Shadow Mapping." In *Proceedings of Graphics Interface*, pp. 147–154. Toronto, Canada: Canadian Information Processing Society, 2008.

[514] M. Schwarzler. "Real-Time Soft Shadows with Adaptive Light Source Sampling." In *Proceedings of the Central European Seminar on Computer Graphics (CESCG)*, 2009. Available online (http://www.cescg.org/proceedings.html).

[515] M. Segal, C. Korobkin, R. van Widenfelt, J. Foran, and P. Haeberli. "Fast Shadow and Lighting Effects Using Texture Mapping." *Proceedings of SIGGRAPH '92, Computer Graphics* 26:2 (1992), 249–252.

[516] B. Segovia, J. Iehl, Richard Mitanchey, and B. Péroche. "Bidirectional Instant Radiosity." In *Eurographics Symposium on Rendering*, pp. 389–397. London: Springer-Verlag, 2006.

[517] B. Segovia, J. Iehl, and B. Péroche. "Metropolis Instant Radiosity." *Computer Graphics Forum (Proceedings of Eurographics)* 26:3 (2007), 425–434.

[518] P. Sen, M. Cammarano, and P. Hanrahan. "Shadow Silhouette Maps." *ACM Transactions on Graphics (Proceedings of SIGGRAPH '03)* 22:3 (2003), 521–526.

[519] S. Seo, D. Kang, Y. Park, and K. Yoon. "The New Area Subdivision and Shadow Generation Algorithms for Colored Paper Mosaic Rendering." *Journal of KCGS* 7:2 (2001), 11–19.

[520] J. Shade, J. Gortler, L. He, and R. Szeliski. "Layered Depth Images." In *Proceedings of SIGGRAPH '98, Computer Graphics Proceedings, Annual Conference Series*, pp. 231–242. New York: ACM, 1998.

[521] P. Shanmugam and O. Arikan. "Hardware Accelerated Ambient Occlusion Techniques on GPUs." In *Proceedings of ACM Symposium on Interactive 3D Graphics and Games*, pp. 73–80. New York: ACM, 2007.

[522] L. Shen, G. Guennebaud, B. Yang, and J. Feng. "Predicted Virtual Soft Shadow Maps with High Quality Filtering." *Computer Graphics Forum (Proceedings of Eurographics)* 30:2 (2011), 493–502.

[523] K. Shimada, A. Yamada, and T. Itoh. "Anisotropic Triangular Meshing of Parametric Surfaces via Close Packing of Ellipsoidal Bubbles." In *Meshing Roundtable, Sandia National Laboratories*, pp. 375–390, 1997.

[524] M. Shinya, T. Takahashi, and S. Naito. "Principles and Applications of Pencil Tracing." *Proceedings of SIGGRAPH '87, Computer Graphics* 21:4 (1987), 45–54.

[525] P. Shirley and S. Marschner. *Fundamentals of Computer Graphics*, Third edition. Natick, MA: A K Peters, 2010.

[526] P. Shirley and R. Morley. *Realistic Ray Tracing*, Second edition. Natick, MA: A K Peters, 2003.

[527] P. Shirley and C. Wang. "Direct Lighting Calculation by Monte Carlo Integration." In *Eurographics Workshop on Rendering*, pp. 54–59. London: Springer-Verlag, 1991.

[528] P. Shirley, C. Wang, and K. Zimmerman. "Monte Carlo Methods for Direct Lighting Calculation." *ACM Transactions on Graphics* 15:1 (1996), 1–36.

[529] P. Shirley. "A Ray Tracing Method for Illumination Calculation in Diffuse-Specular Scenes." In *Proceedings of Graphics Interface*, pp. 205–212. Toronto, Canada: Canadian Information Processing Society, 1990.

[530] F. Sillion and C. Puech. *Radiosity and Global Illumination*. San Francisco: Morgan Kaufmann, 1994.

[531] C. Silva, Y. Chiang, J. El-Sana, and P. Lindstrom. "Out-of-Core Algorithms for Scientific Visualization and Computer Graphics." In *IEEE Visualization (Courses)*. Washington, DC: IEEE Computer Society, 2002.

[532] J. Singh and P. Narayanan. "Real-Time Ray-Tracing of Implicit Surfaces on the GPU." *IEEE Transactions on Visualization and Computer Graphics* 16:2 (2009), 261–272.

[533] E. Sintorn and U. Assarsson. "Real-Time Approximate Sorting for Self-Shadowing and Transparency in Hair Rendering." In *Proceedings of ACM Symposium on Interactive 3D Graphics and Games*. New York: ACM, 2008.

[534] E. Sintorn and U. Assarsson. "Hair Self Shadowing and Transparency Depth Ordering Using Occupancy Maps." In *Proceedings of ACM Symposium on Interactive 3D Graphics and Games*, pp. 67–74. New York: ACM, 2009.

[535] E. Sintorn, E. Eisemann, and U. Assarsson. "Sample based Visibility for Soft Shadows Using Alias-Free Shadow Maps." *Computer Graphics Forum (Proceedings of Eurographics)* 27:4 (2008), 1285–1292.

[536] E. Sintorn, O. Olsson, and U. Assarsson. "An Efficient Alias-Free Shadow Algorithm for Opaque and Transparent Objects Using Per-Triangle Shadow Volumes." *ACM Transactions on Graphics (Proceedings of SIGGRAPH Asia '11)* 30:6 (2011), 153:1–10.

[537] M. Slater. "A Comparison of Three Shadow Volume Algorithms." *The Visual Computer* 9:1 (1992), 25–38.

[538] P.-P. Sloan and M. Cohen. "Interactive Horizon Mapping." In *Eurographics Workshop on Rendering*, pp. 281–286. London: Springer-Verlag, 2000.

[539] P.-P. Sloan, J. Kautz, and J. Snyder. "Precomputed Radiance Transfer for Real-Time Rendering in Dynamic, Low-Frequency Lighting Environments." *ACM Transactions on Graphics (Proceedings of SIGGRAPH '02)* 21:3 (2002), 527–536.

[540] P.-P. Sloan, J. Hall, J. Hart, and J. Snyder. "Clustered Principal Components for Pre-computed Radiance Transfer." *ACM Transactions on Graphics (Proceedings of SIGGRAPH '03)* 22:3 (2003), 382–391.

[541] P.-P. Sloan, N.K. Govindaraju, D. Nowrouzezahrai, and J. Snyder. "Image-Based Proxy Accumulation for Real-Time Soft Global Illumination." In *Proceedings of Pacific Graphics*, pp. 97–105. Washington, DC: IEEE Computer Society, 2007.

[542] M. Slomp, M. Oliveira, and D. Patricio. "A Gentle Introduction to Precomputed Radiance Transfer." *RITA* 13:2 (2006), 131–160.

[543] B.G. Smith. "Geometrical Shadowing of a Random Rough Surface." *IEEE Transactions on Antennas and Propagation* AP-15:5 (1967), 668–671.

[544] B. Smits, J. Arvo, and D. Greenberg. "A Clustering Algorithm for Radiosity in Complex Environments." In *Proceedings of SIGGRAPH '94, Computer Graphics Proceedings, Annual Conference Series*, pp. 435–442. New York: ACM, 1994.

[545] B. Smits. "Efficiency Issues for Ray Tracing." *Journal of Graphics, GPU, and Game Tools* 3:2 (1998), 1–14.

[546] J. Snyder and A. Barr. "Ray Tracing Complex Models Containing Surface Tessellations." *Proceedings of SIGGRAPH '87, Computer Graphics* 21:4 (1987), 119–128.

[547] J. Snyder and D. Nowrouzezahrai. "Fast Soft Self-Shadowing on Dynamic Height Fields." *Computer Graphics Forum (Eurographics Symposium on Rendering)* 27:4 (2008), 1275–1283.

[548] J. Snyder, R. Barzel, and S. Gabriel. "Motion Blur on Graphics Workstations." In *Graphics Gems III*, pp. 374–382. New York: Academic Press, 1992.

[549] L. Sobierajski and R. Avila. "A Hardware Acceleration Method for Volumetric Ray Tracing." In *Proceedings of IEEE Visualization*, pp. 27–34. Washington, DC: IEEE Computer Society, 1995.

[550] C. Soler and F. Sillion. "Fast Calculation of Soft Shadow Textures Using Convolution." In *Proceedings of SIGGRAPH '98, Computer Graphics Proceedings, Annual Conference Series*, pp. 321–332. New York: ACM, 1998.

[551] C. Soler and F. Sillion. "Texture-Based Visibility for Efficient Lighting Simulation." In *Eurographics Workshop on Rendering*, pp. 199–210. London: Springer-Verlag, 1998.

[552] M. Soucy and D. Laurendeau. "A General Surface Approach to the Integration of a Set of Range Views." *IEEE Transactions on Pattern Analysis and Machine Intelligence* 17:4 (1995), 344–358.

[553] M. Sousa and J. Buchanan. "Computer-Generated Graphite Pencil Rendering of 3D Polygonal Models." *Computer Graphics Forum* 18:3 (1999), 195–208.

[554] M. Spindler, N. Rober, R. Dohring, and M. Masuch. "Enhanced Cartoon and Comic Rendering." In *Eurographics (Short Papers)*, pp. 141–144. Aire-la-Ville, Switzerland: Eurographics Association, 2006.

[555] J.-F. St-Amour, E. Paquette, and P. Poulin. "Soft Shadows from Extended Light Sources with Deep Shadow Maps." In *Proceedings of Graphics Interface*, pp. 105–112. Toronto, Canada: Canadian Information Processing Society, 2005.

[556] J. Stam. "Diffraction Shaders." In *Proceedings of SIGGRAPH '99, Computer Graphics Proceedings, Annual Conference Series*, pp. 101–110. New York: ACM, 1999.

[557] M. Stamminger and G. Drettakis. "Perspective Shadow Maps." *ACM Transactions on Graphics (Proceedings of SIGGRAPH '02)* 21:3 (2002), 557–562.

[558] M. Stark, E. Cohen, T. Lyche, and R. Fiesenfeld. "Computing Exact Shadow Irradiance Using Splines." In *Proceedings of SIGGRAPH '99, Computer Graphics Proceedings, Annual Conference Series*, pp. 155–164. New York: ACM, 1999.

[559] A. State, G. Hirota, D. Chen, W. Garrett, and M. Livingston. "Superior Augmented Reality Registration by Integrating Landmark Tracking and Magnetic Tracking." In *Proceedings of SIGGRAPH '96, Computer Graphics Proceedings, Annual Conference Series*, pp. 429–438. New York: ACM, 1996.

[560] J. Stewart and S. Ghali. "An Output Sensitive Algorithm for the Computation of Shadow Boundaries." In *Canadian Conference on Computational Geometry*, pp. 291–296, 1993.

[561] J. Stewart and S. Ghali. "Fast Computation of Shadow Boundaries Using Spatial Coherence and Backprojections." In *Proceedings of SIGGRAPH '94, Computer Graphics Proceedings, Annual Conference Series*, pp. 231–238. New York: ACM, 1994.

[562] J. Stewart. "Fast Horizon Computation at all Points of a Terrain with Visibility and Shading Applications." *IEEE Transactions on Visualization and Computer Graphics* 4:1 (1998), 82–93.

[563] J. Stewart. "Computing Visibility from Folded Surfaces." *Computers and Graphics* 23:5 (1999), 693–702.

[564] A.J. Stewart. "Vicinity Shading for Enhanced Perception of Volumetric Data." In *Proceedings of IEEE Visualization*, pp. 355–362. Washington, DC: IEEE Computer Society, 2003.

[565] M. Stich, C. Wachter, and A. Keller. "Efficient and Robust Shadow Volumes Using Hierarchical Occlusion Culling and Geometry Shaders." In *GPU Gems 3*, pp. 239–256. Reading, MA: Addison Wesley, 2007.

[566] J. Stoker. "Voxels as a Representation of Multiple-return LIDAR Data." In *Proceedings of ASPRS*, 2004.

[567] S. Sudarsky. "Generating Dynamic Shadows for Virtual Reality Applications." In *Proceedings of the International Conference on Information Visualization*, p. 595. Washington, DC: IEEE Computer Society, 2001.

[568] N. Sugano, H. Kato, and K. Tachibana. "The Effects of Shadow Representation of Virtual Objects in Augmented Reality." In *Proceedings of International Symposium on Mixed and Augmented Reality (ISMAR)*, pp. 76–83. Los Alamitos, CA: IEEE, 2003.

[569] K. Sung, A. Pearce, and C. Wang. "Spatial Temporal Anti-aliasing." *IEEE Transactions on Visualization and Computer Graphics* 8:2 (2002), 144–153.

[570] K. Sung. "Area Sampling Buffer: Tracing Rays with Z-buffer Hardware." *Computer Graphics Forum (Proceedings of Eurographics)* 11:3 (1992), 299–310.

[571] P. Supan, I. Stuppacher, and M. Haller. "Image based Shadowing in Real-Time Augmented Reality." *International Journal of Virtual Reality* 5:3 (2006), 1–7.

[572] M. Sweeney and R. Bartels. "Ray Tracing Free-form B-Spline Surfaces." *IEEE Computer Graphics and Applications* 6:2 (1986), 41–49.

[573] L. Szirmay-Kalos and T. Umenhoffer. "Displacement Mapping on the GPU—State-of-the-Art." *Computer Graphics Forum* 27:6 (2008), 1567–1592.

[574] L. Szirmay-Kalos, T. Umenhoffer, B. Toth, L. Szecsi, and M. Sbert. "Volumetric Ambient Occlusion." *IEEE Computer Graphics and Applications* 30:1 (2010), 70–79.

[575] K. Tadamura, X. Qin, G. Jiao, and E. Nakamae. "Rendering Optimal Solar Shadows with Plural Sunlight Depth Buffers." *The Visual Computer* 17:2 (2001), 76–90.

[576] T. Tanaka and T. Takahashi. "Cross Scan Buffer and Its Applications." *Computer Graphics Forum (Proceedings of Eurographics)* 13:3 (1994), 467–476.

[577] T. Tanaka and T. Takahashi. "Fast Shadowing Algorithm for Linear Light Sources." *Computer Graphics Forum (Proceedings of Eurographics)* 14:3 (1995), 205–216.

[578] T. Tanaka and T. Takahashi. "Fast Analytic Shading and Shadowing for Area Light Sources." *Computer Graphics Forum (Proceedings of Eurographics)* 16:3 (1997), 231–240.

[579] M. Tang and J.-X. Dong. "Geometry Image-Based Shadow Volume Algorithm for Subdivision Surfaces." In *Proceedings of Computer Graphics International*, pp. 21–28. Washington, DC: IEEE Computer Society, 2007.

[580] M. Tang, J.-X. Dong, and S.-C. Chou. "Real-Time Shadow Volumes for Subdivision Surface based Models." In *Proceedings of Computer Graphics International*, pp. 538–545. Washington, DC: IEEE Computer Society, 2006.

[581] N. Tatarchuk. "Dynamic Parallax Occlusion Mapping with Approximate Soft Shadows." In *Proceedings of ACM Symposium on Interactive 3D Graphics and Games*, pp. 63–69. New York: ACM, 2006.

[582] N. Tatarchuk. "Practical Parallax Occlusion Mapping with Approximate Soft Shadows for Detailed Surface Rendering." In *ShaderX⁵: Advanced Rendering Techniques*, pp. 75–105. Hingham, MA: Charles River Media, 2006.

[583] F. Tecchia, C. Loscos, and Y. Chrysanthou. "Visualizing Crowds in Real-Time." *Computer Graphics Forum* 21:4 (2002), 753–765.

[584] S. Teller. "Computing the Antipenumbra of an Area Light Source." *Proceedings of SIGGRAPH '92, Computer Graphics* 26:2 (1992), 139–148.

[585] K. Thakur, F. Cheng, and K. Miura. "Shadow Generation Using Discretized Shadow Volume in Angular Coordinates." In *Pacific Conference on Computer Graphics and Applications*, pp. 224–233. Washington, DC: IEEE Computer Society, 2003.

[586] V. Timonen and J. Westerholm. "Scalable Height Field Self-Shadowing." *Computer Graphics Forum (Proceedings of Eurographics)* 29:2 (2010), 723–731.

[587] O. Tolba, J. Dorsey, and L. McMillan. "A Projective Drawing System." In *Proceedings of ACM Symposium on Interactive 3D Graphics*, pp. 25–34. New York: ACM, 2001.

[588] R. Toledo, B. Levy, and J. Paul. "Iterative Methods for Visualization of Implicit Surfaces on GPU." In *Advances in Visual Computing*, pp. 598–609. Berlin: Springer-Verlag, 2007.

[589] K.E. Torrance and E.M. Sparrow. "Theory for Off-Specular Reflection from Roughened Surfaces." *Journal of Optical Society of America* 57:9 (1967), 1105–1112.

[590] B. Toth and T. Umenhoffer. "Real-Time Volumetric Lighting in Participating Media." In *Eurographics (Short Papers)*. Aire-la-Ville, Switzerland: Eurographics Association, 2009.

[591] D. Toth. "On Ray Tracing Parametric Surfaces." *Proceedings of SIGGRAPH '85, Computer Graphics* 19:3 (1985), 171–179.

[592] T.S. Trowbridge and K.P. Reitz. "Average Irregularity Representation of a Roughened Surface for Ray Reflection." *Journal of the Optical Society of America* 65:5 (1975), 531–536.

[593] Y.-T. Tsai and Z.-C. Shih. "All-Frequency Precomputed Radiance Transfer Using Spherical Radial Basis Functions and Clustered Tensor Approximation." *ACM Transactions on Graphics (Proceedings of SIGGRAPH '06)* 25:3 (2006), 967–976.

[594] G. Turk and M. Levoy. "Zippered Polygon Meshes from Range Images." In *Proceedings of SIGGRAPH '94, Computer Graphics Proceedings, Annual Conference Series*, pp. 311–318. New York: ACM, 1994.

[595] T. Udeshi and C. Hansen. "Towards Interactive, Photorealistic Rendering of Indoor Scenes: A Hybrid Approach." In *Eurographics Workshop on Rendering*, pp. 63–76. London: Springer-Verlag, 1999.

[596] Y. Uralsky. "Efficient Soft-edged Shadows Using Pixel Shader Branching." In *GPU Gems* 2, pp. 269–282. Reading, MA: Addison Wesley, 2005.

[597] M. Valient and W. deBoer. "Fractional-Disk Soft Shadows." In *ShaderX⁴: Advanced Rendering Techniques*, pp. 411–423. Hingham, MA: Charles River Media, 2004.

[598] M. Valient. "Deferred Rendering in Killzone 2." Presented at Develop Conference, 2007.

[599] M. Valient. "Stable Rendering of Cascaded Shadow Maps." In *ShaderX⁶: Advanced Rendering Techniques*, pp. 231–238. Hingham, MA: Charles River Media, 2008.

[600] J. van Ee and C. van Overveld. "Casting Shadows with Approximated Object Space Accuracy by Means of a Modified Z-buffer." *The Visual Computer* 10:5 (1994), 243–254.

[601] J. van Waveren. "Shadow Volume Construction." In *Intel Software Network*, 2005. Available online (http://software.intel.com/file/37730).

[602] E. Veach and L. Guibas. "Metropolis Light Transport." In *Proceedings of Computer Graphics and Interactive Techniques*, pp. 65–76. New York: ACM, 1997.

[603] C. Vedel. "Computing Illumination from Area Light Sources by Approximate Contour Integration." In *Proceedings of Graphics Interface*, pp. 237–243. Toronto, Canada: Canadian Information Processing Society, 1993.

[604] R. Viney. "Algorithms for Real-Time Rendering of Soft Shadows." Technical report, University of Canterbury, 2007.

[605] A. Vlachos and D. Card. "Computing Optimized Shadow Volumes for Complex Data Sets." In *Game Programming Gems 3*, pp. 367–371. Hingham, MA: Charles River Media, 2002.

[606] A. Vlachos, D. Gosselin, and J. Mitchell. "Self-Shadowing Characters." In *Game Programming Gems II*, pp. 220–227. New York: Academic Press, 2001.

[607] V. Vlassopoulos. "Adaptive Polygonization of Parametric Surfaces." *The Visual Computer* 6:5 (1990), 291–298.

[608] I. Wald and H. P. Seidel. "Interactive Ray Tracing of Point-Based Models." In *Proceedings of Eurographics Symposium on Point-Based Graphics*, pp. 1–8. Los Alamitos, CA: IEEE, 2005.

[609] I. Wald, P. Slusallek, and C. Benthin. "Interactive Distributed Ray Tracing of Highly Complex Models." In *Eurographics Workshop on Rendering*, pp. 277–288. London: Springer-Verlag, 2001.

[610] I. Wald, T. Purcell, J. Schmittler, C. Benthin, and P. Slusallek. "Realtime Ray Tracing and Its Use for Interactive Global Illumination." In *State-of-the-Art Report in Proceedings of Eurographics*, p. 24. Aire-la-Ville, Switzerland: Eurographics, 2003.

[611] I. Wald, A. Dietrich, and P. Slusallek. "An Interactive Out-of-Core Rendering Framework for Visualizing Massively Complex Models." In *Eurographics Symposium on Rendering*, pp. 81–92. Aire-la-Ville, Switzerland: Eurographics Association, 2004.

[612] I. Wald, S. Boulos, and P. Shirley. "Ray Tracing Deformable Scenes Using Dynamic Bounding Volume Hierarchies." *ACM Transactions on Graphics* 26:1 (2007), 6:1–6.

[613] I. Wald, W. Mark, J. Gunther, S. Boulos, T. Ize, W. Hunt, S. Parker, and P. Shirley. "State-of-the-Art in Ray Tracing Animated Scenes." In *State-of-the-Art Report in Proceedings of Eurographics*. Aire-la-Ville, Switzerland: Eurographics, 2007.

[614] B. Walter, S. Fernandez, A. Arbree, K. Bala, M. Donikian, and D.P. Greenberg. "Lightcuts: A Scalable Approach to Illumination." *ACM Transactions on Graphics (Proceedings of SIGGRAPH '05)* 24:3 (2005), 1098–1107.

[615] M. Wand and W. Strasser. "Multi-Resolution Point Sample Raytracing." In *Proceedings of Graphics Interface*, pp. 139–148. Toronto, Canada: Canadian Information Processing Society, 2003.

[616] Y. Wang and S. Molnar. "Second-Depth Shadow Mapping." Technical Report TR94-019, Computer Science, University of North Carolina, Chapel Hill, 1994.

[617] C. Wang and K. Sung. "Multi-stage N-Rooks Sampling Method." *Journal of Graphics, GPU, and Game Tools* 4:1 (1999), 39–47.

[618] T. Wang, P. Sheu, and S. Hwang. "An Object-Oriented Shadow Generation Algorithm for Real-Time Application." In *Workshop on Object-Oriented Real-Time Dependable Systems*, pp. 17–25. Washington, DC: IEEE Computer Society, 1997.

[619] L. Wang, X. Wang, X. Tong, S. Lin, S. Hu, B. Guo, and H. Shum. "View-Dependent Displacement Mapping." *ACM Transactions on Graphics (Proceedings of SIGGRAPH '03)* 22:3 (2003), 334–339.

[620] X. Wang, X. Tong, S. Lin, S. Hu, B. Guo, and H. Shum. "Generalized Displacement Maps." In *Eurographics Symposium on Rendering*, pp. 227–233. Aire-la-Ville, Switzerland: Eurographics Association, 2004.

[621] C. Wang. "Physically Correct Direct Lighting for Distribution Ray Tracing." In *Graphics Gems III*, pp. 303–313. New York: Academic Press, 1992.

[622] L. Wanger, J. Ferwerda, and D. Greenberg. "Perceiving Spatial Relationships in Computer Generated Images." *IEEE Computer Graphics and Applications* 12:3 (1992), 44–55.

[623] L. Wanger. "The Effect of Shadow Quality on the Perception of Spatial Relationships in Computer Generated Imagery." In *Proceedings of ACM Symposium on Interactive 3D Graphics*, pp. 39–42. New York: ACM, 1992.

[624] G. Ward, F. Rubinstein, and R. Clear. "A Ray Tracing Solution for Diffuse Interreflection." *Proceedings of SIGGRAPH '88, Computer Graphics* 22:4 (1988), 85–92.

[625] G. Ward. "Adaptive Shadow Testing for Ray Tracing Photorealistic Rendering in Computer Graphics." In *Eurographics Workshop on Rendering*. London: Springer-Verlag, 1991.

[626] A. Watt and F. Policarpo. "Relief Maps with Silhouettes." In *Advanced Game Development with Programmable Graphics Hardware*. Wellesley, MA: A K Peters, 2005.

[627] D. Weiskopf and T. Ertl. "Shadow Mapping Based on Dual Depth Layers." In *Eurographics (Short Papers)*, pp. 53–60. Aire-la-Ville, Switzerland: Eurographics Association, 2003.

[628] L. Westover. "Footprint Evaluation for volume Rendering." *Proceedings of SIGGRAPH '90, Computer Graphics* 24:4 (1990), 367–376.

[629] T. Whitted. "An Improved Illumination Model for Shaded Display." *Communications of the ACM* 23:6 (1980), 343–349.

[630] L. Williams. "Casting Curved Shadows on Curved Surfaces." In *Proceedings of SIGGRAPH '78*, pp. 270–274. New York: ACM, 1978.

[631] M. Wimmer and J. Bittner. "Hardware Occlusion Queries Made Useful." In *GPU Gems 2*, pp. 91–108. Reading, MA: Addison Wesley, 2005.

[632] M. Wimmer and D. Scherzer. "Robust Shadow Mapping with Light-Space Perspective Shadow Mapping." In *Shader⁴: Advanced Rendering Techniques*, pp. 313–330. Hingham, MA: Charles River Media, 2006.

[633] M. Wimmer, D. Scherzer, and W. Purgathofer. "Light Space Perspective Shadow Maps." In *Eurographics Symposium on Rendering*, pp. 143–152. Aire-la-Ville, Switzerland: Eurographics Association, 2004.

[634] G. Winkenbach and D. Salesin. "Computer-Generated Pen-and-ink Illustration." In *Proceedings of SIGGRAPH '94, Computer Graphics Proceedings, Annual Conference Series*. New York: ACM, 1994.

[635] G. Winkenbach and D. Salesin. "Rendering Parametric Surfaces in Pen and Ink." In *Proceedings of SIGGRAPH '96, Computer Graphics Proceedings, Annual Conference Series*, pp. 469–476. New York: ACM, 1996.

[636] K. Wong and W. Tsang. "An Efficient Shadow Algorithm for Area Light Sources Using BSP Trees." In *Proceedings of Pacific Graphics*, pp. 97–102. Washington, DC: IEEE Computer Society, 1998.

[637] T. Wong, W. Luk, and P. Heng. "Sampling with Hammersley and Halton Points." *Journal of Graphics, GPU, and Game Tools* 2:2 (1997), 9–24.

[638] A. Woo and J. Amanatides. "Voxel Occlusion Testing: A Shadow Determination Accelerator for Ray Tracing." In *Proceedings of Graphics Interface*, pp. 213–220. Toronto, Canada: Canadian Information Processing Society, 1990.

[639] A. Woo, P. Poulin, and A. Fournier. "A Survey of Shadow Algorithms." *IEEE Computer Graphics and Applications* 10:6 (1990), 13–32.

[640] A. Woo, M. Ouellette, and A. Pearce. "It Is Really Not a Rendering Bug, You See..." *IEEE Computer Graphics and Applications* 16:5 (1996), 21–25.

[641] A. Woo. "Ray Tracing Polygons Using Spatial Subdivision." In *Proceedings of Graphics Interface*, pp. 184–191. Toronto, Canada: Canadian Information Processing Society, 1992.

[642] A. Woo. "The Shadow Depth Map Revisited." In *Graphics Gems III*, pp. 338–342. New York: Academic Press, 1992.

[643] A. Woo. "Efficient Shadow Computations in Ray Tracing." *IEEE Computer Graphics and Applications* 15:3 (1993), 78–83.

[644] C. Woodward. "Ray Tracing Parametric Surfaces by Subdivision in Viewing Plane." In *Theory and Practice of Geometric Modeling*, edited by Wolfgang Straßer and Hans-Peter Seidel, pp. 273–287. New York: Springer-Verlag, 1989.

[645] M. Wrenninge, N. Bin Zafar, J. Clifford, G. Graham, D. Penney, J. Kontkanen, J. Tessendorf, and A. Clinton. "Efficient Image-Based Methods for Rendering Soft Shadows." In *SIGGRAPH (Courses)*, 2010.

[646] C. Wyman and C. Hansen. "Penumbra Maps: Approximate Soft Shadows in Real-Time." In *Eurographics Symposium on Rendering*, pp. 202–207. Aire-la-Ville, Switzerland: Eurographics Association, 2003.

[647] C. Wyman and S. Ramsey. "Interactive Volumetric Shadows in Participating Media with Single-Scattering." In *Proceedings of IEEE Symposium on Interactive Ray Tracing*, pp. 87–92. Washington, DC: IEEE Computer Society, 2008.

[648] C. Wyman. "Interactive Voxelized Epipolar Shadow Volumes." In *SIGGRAPH Asia (Sketches)*, pp. 53:1–2. New York: ACM, 2010.

[649] F. Xie, E. Tabellion, and A. Pearce. "Soft Shadows by Ray Tracing Multilayer Transparent Shadow Maps." In *Eurographics Symposium on Rendering*, pp. 265–276. Aire-la-Ville, Switzerland: Eurographics Association, 2007.

[650] R. Yagel, D. Cohen, and A. Kaufman. "Discrete Ray Tracing." *IEEE Computer Graphics and Applications* 12:5 (1992), 19–28.

[651] B. Yang, J. Feng, G. Guennebaud, and X. Liu. "Packet-Based Hierarchical Soft Shadow Mapping." *Computer Graphics Forum (Eurographics Symposium on Rendering)* 28:4 (2009), 1121–1130.

[652] Z. Ying, M. Tang, and J. Dong. "Soft Shadow Maps for Area Light by Area Approximation." In *Pacific Conference on Computer Graphics and Applications*, pp. 442–443. Washington, DC: IEEE Computer Society, 2002.

[653] K.-H. Yoo, D. Kim, S. Shin, and K.-Y. Chwa. "Linear-Time Algorithms for Finding the Shadow Volumes from a Convex Area Light Source." *Algorithmica* 20:3 (1998), 227–241.

[654] K. Yoo, D. Kim, S. Shin, and K. Chwa. "Linear-Time Algorithms for Finding the Shadow Volumes from a Convex Area Light Source." *Algorithmica Journal* 20:3 (1998), 227–241.

[655] C. Yuksel and J. Keyser. "Deep Opacity Maps." *Computer Graphics Forum (Proceedings of Eurographics)* 27:2 (2008), 675–680.

[656] H. Zatz. "Galerkin Radiosity: A Higher Order Solution Method for Global Illumination." In *Proceedings of SIGGRAPH '93, Computer Graphics Proceedings, Annual Conference Series*, pp. 213–220. New York: ACM, 1993.

[657] C. Zhang and R. Crawfis. "Volumetric Shadows Using Splatting." In *Proceedings of IEEE Visualization*, pp. 85–92. Washington, DC: IEEE Computer Society, 2002.

[658] C. Zhang and R. Crawfis. "Shadows and Soft Shadows with Participating Media Using Splatting." *IEEE Transactions on Visualization and Computer Graphics* 9:2 (2003), 139–149.

[659] X. Zhang and M. Nakajima. "Image-Based Building Shadow Generation Technique for Virtual Outdoor Scene." In *Proceedings of International Conference on Artificial Reality and Telexistence (ICAT)*, pp. 132–139, 2000.

[660] X. Zhang and M. Nakajima. "Image-Based Tree Shadow Morphing Technique." *Journal of the Institute of Image Electronics Engineers of Japan* 29:5 (2000), 553–560.

[661] X. Zhang and M. Nakajima. "A Study on Image-Based Scene Modeling and Rendering—Part IV : Building Shadow Generation by Shape from Shadow Silhouette." In *Information and Systems Society Conference of IEICE*, D-12-63, 2000.

[662] C. Zhang, D. Xue, and R. Crawfis. "Light Propagation for Mixed Polygonal and Volumetric Data." In *Proceedings of Computer Graphics International*, pp. 249–256. Washington, DC: IEEE Computer Society, 2005.

[663] F. Zhang, H. Sun, L. Xu, and K. L. Lee. "Parallel-Split Shadow Maps for Large-Scale Virtual Environments." In *ACM International Conference on Virtual Reality Continuum and Its Applications*, pp. 311–318, 2006.

[664] F. Zhang, H. Sun, L. Xu, and L. Lun. "Parallel-Split Shadow Maps for Large-Scale Virtual Environments." In *ACM International Conference on Virtual Reality Continuum and Its Applications (VCRIA)*, pp. 311–318. New York: ACM, 2006.

[665] F. Zhang, L. Xu, C. Tao, and H. Sun. "Generalized Linear Perspective Shadow Map Reparameterization." In *ACM International Conference on Virtual Reality Continuum and Its Applications (VCRIA)*, pp. 339–342. New York: ACM, 2006.

[666] Y. Zhang, O. Teboul, X. Zhang, and Q. Deng. "Image based Real-Time and Realistic Forest Rendering and Forest growth Simulation." In *Proceedings of International Symposium on Plant Growth Modeling, Simulation, Visualization and Applications (PMA)*, pp. 323–327. Washington, DC: IEEE Computer Society, 2007.

[667] F. Zhang, H. Sun, L. Xu, and L. Lun. "Hardware-Accelerated Parallel-Split Shadow Maps." *International Journal of Image and Graphics* 8:2 (2008), 223–241.

[668] F. Zhang, A. Zaprjagaev, and A. Bentham. "Practical Cascaded Shadow Mapping." In *ShaderX⁷: Advanced Rendering Techniques*, pp. 305–330. Hingham, MA: Charles River Media, 2009.

[669] H. Zhang. "Forward Shadow Mapping." In *Eurographics Workshop on Rendering*, pp. 131–138. London: Springer-Verlag, 1998.

[670] F. Zhang. "Theory to Practice: Generalized Minimum-norm Perspective Shadow Maps for Anti-aliased Shadow Rendering in 3D Computer Games." *Lecture Notes in Computer Science* 4740 (2007), 66–78.

[671] N. Zhao, Y. Chen, and Z. Pan. "Antialiased Shadow Algorithms for Game Rendering." *Proceedings of International Conference on Edutainment* 3942 (2006), 873–882.

[672] Z. Zheng and S. Saito. "Screen Space Anisotropic Blurred Soft Shadows." In *SIGGRAPH (Sketches)*, p. 1. New York: ACM, 2011.

[673] K. Zhou, Y. Hu, S. Lin, B. Guo, and H. Shum. "Precomputed Shadow Fields for Dynamic Scenes." *ACM Transactions on Graphics (Proceedings of SIGGRAPH '05)* 24:3 (2005), 1196–1201.

[674] S. Zhukov, A. Inoes, and G. Kronin. "An Ambient Light Illumination Model." In *Eurographics Workshop on Rendering*, pp. 45–56. London: Springer-Verlag, 1998.

[675] R. Zioma. "Reverse Extruded Shadow Volumes." In *ShaderX²: Shader Programming Tips and Tricks*, pp. 587–593. Hingham, MA: Charles River Media, 2003.

[676] M. Zlatuska and V. Havran. "Ray Tracing on a GPU with CUDA: Comparative Study of Three Algorithms." *Journal of WSCG* 18:1–3 (2010), 69–76.

Index

T - #0411 - 071024 - C268 - 229/152/12 - PB - 9780367381240 - Gloss Lamination